D0937277

THE ENGINEERING PROJECT

THE ENGINEERING PROJECT

Its Nature, Ethics, and Promise

GENE MORIARTY

THE PENNSYLVANIA STATE UNIVERSITY PRESS
UNIVERSITY PARK, PENNSYLVANIA

LIBRARY OF CONGRESS
CATALOGING-IN-PUBLICATION DATA

Moriarty, Gene.
The engineering project : its nature, ethics, and promise / Gene Moriarty.
p. cm.
Includes bibliographical references and index.
ISBN 978-0-271-03254-2 (cloth : alk. paper)
1. Engineering.
I. Title.

TA157.M835 2008
620—dc22
2007031897

Copyright © 2008
The Pennsylvania State University
All rights reserved
Printed in the United States of America
Published by
The Pennsylvania State University Press,
University Park, PA 16802-1003

The Pennsylvania State University Press is a member
of the Association of American University Presses.

It is the policy of The Pennsylvania State University
Press to use acid-free paper. This book is printed on
Natures Natural, containing 50% post-consumer waste,
and meets the minimum requirements of American
National Standard for Information Sciences—
Permanence of Paper for Printed Library Material,
ANSI Z39.48–1992.

CONTENTS

PREFACE AND ACKNOWLEDGMENTS

I have been a faculty member in the Electrical Engineering Department at San Jose State University for many years. My technical areas of focus have been circuits, systems, and controls, but while finishing my dissertation in electrical engineering in the 1970s, I discovered philosophy, in particular, the nascent field of philosophy of technology. That early interest has continued and developed over the years—with an emphasis on the philosophical aspects of engineering ethics—and has led to my development of the ideas and material presented in this book.

I am deeply indebted to Albert Borgmann for many years of inspiration. One of my fondest memories is, while on sabbatical from San Jose State in the middle 1980s, spending a couple of months sitting in his course at the University of Montana in Missoula. From his work I have learned much. Carl Mitcham also stands out as a philosopher of technology from whom I have profited immensely, as do Langdon Winner, Andrew Feenberg, and Don Ihde. Feedback from friends and colleagues over the years has been invaluable. Among them are Mischa Adams, Harriet Brown, Stephen Brown, Jan Browne, Guy Cox, Jim Freeman, Chris Gaynor, Rick Gladstone, Fran Guerrero, Ken Harmon, Joe Herkert, Laura Hughes, Barbara Joans, John Lamandella, Kara Lindstrom, Pete Shanks, Lowell Smith, Michael Tanner, Vito Victor, and Kam Yuen. The editing work of Andrew B. Lewis has been of tremendous help to this project.

Much of my work, at various stages of development, has been previously presented in the following essays:

"Ethics, Ethos and the Professions: Some Lessons from Engineering." *Professional Ethics* 4, no. 1 (1995).

"Engineering Design: Content and Context." *Journal of Engineering Education* 83, no. 2 (1994).

"Engineering Education in the Postmodern Era." *Frontiers in Education Conference Proceedings* (1994), FIE Conference in San Jose, Calif.

"The Place of Focal Engineering in University Education," *International Symposium on Technology and Society Conference Proceedings* (2000), ISTAS Conference in Rome.

"A Plea for the Practice of Focal Engineering." *Technology in Society* 21, no. 4 (1999).

"Technology: The Cost of Disengagement/The Benefit of Disburdenment," *International Symposium on Technology and Society Conference Proceedings* (1993), ISTAS Conference in Washington, D.C.

"Three Kinds of Ethics for Three Kinds of Engineering." *IEEE Technology and Society Magazine* 20, no. 3 (2001).

INTRODUCTION

Engineering is the practice of making good on the promise of technology. Technology, throughout history, has promised relief from the burdens of everyday life. Engineering practice has brought us an array of time- and labor-saving devices. The telephone, for instance, lifts the burden of distance between friends, family, neighbors, and others. In characterizing technology as *disburdening,* philosopher of technology Albert Borgmann also points to its *disengaging* character, which implies that typically we need have only minimal connection to or involvement with engineered devices, and these in turn have minimal connection to or involvement with the worlds in which they are functioning.[1] Disengagement and disburdenment tend to go hand in hand in a world under the influence of modern engineering practice.

Engaging practices and products on the other hand tend to be *burdensome.* Cultivating my vegetable garden is an engaging pursuit. It is also hard work. It is a bit easier if I use a rake and a hoe, instead of, say, my bare hands, which would be highly engaging but quite burdensome. More and more garden tools would reduce my burden, but increase my disengagement. The ultimate disburdenment might be to just shop for my vegetables at the supermarket and be done with it. Somewhere there must be a balance: how much disburdenment will still permit an engagement that enriches life and elevates the spirit? The kind of engineering that contributes to such a balance I call *focal engineering.*

Still, you might ask, is not disburdenment an inherent good? It is certainly good to be relieved from the burden of an onerous task, and it is

1. Albert Borgmann, *Technology and the Character of Contemporary Life* (Chicago: University of Chicago Press, 1984), 33–48.

not necessarily bad to be disengaged from the world in which that onerous task was performed. Without a telephone, a conversation with a friend three hours away on foot would require a six-hour round trip. That would pretty much do in the day. Now I just pick up the phone and am disburdened from all that travail. But along with the disburdenment I am also disengaged from the walk in the woods between his house and mine, an outing would have embellished my being, cleared the cobwebs from my brain, and given me some much needed exercise. However, disengagement is not necessarily problematic. I can still take an engaging walk in the woods after my phone call, or these days I can make the call on my cell phone from wherever I happen to be. In principle, the lifting of burdens frees up the human spirit for other ventures. In fact, thanks to the engineered product of the telephone, compared to life in the pre-telephone era, I now have more hours in my day to do other things.

I am certainly grateful for burdens lifted and time saved. How do I use that time? That is of course a personal matter. I may squander it or spend it wisely. Personally, I tend to spend some of it being entertained by other engineered devices, some of it involving myself in other practices I see as engaging, and some of it attending to a variety of other burdens a given engineered product might engender. My car, for instance, is in fact engaging to drive. It gives me tremendous mobility and pleasure, but at the cost of keeping it properly fueled and maintained. Good engineering minimizes the time needed to address many of those additional burdens. It is surely the case that a well-engineered automobile needs less time in the shop than a poorly engineered one.

Suppose I choose to take my home off the energy grid. Renewable electric power systems usually employ lead acid battery storage and solar cell or photovoltaic panels, which are often augmented by small generators powered by wind or running water.[2] Certain practices are required in order to keep the system together. Unlike standard electrical power, which I can get with the flip of a switch, home power is not so convenient. After a home power system is installed, the real work begins. The battery terminals must be kept clean, and the homeowner must continually adjust energy use to resource availability (sun, wind, water flow).[3] Standard electrical power available from the grid disburdens my life but at the same time

2. Jesse S. Tatum, "Technology and Values: Getting Beyond the 'Device Paradigm' Impasse," *Science, Technology, and Human Values* 19 (Winter 1994): 76.

3. Ibid., 79–80.

tends to disengage me. I may not know or even care where the power comes from. It is just there. I am not just anonymous when I draw power from the grid, I can be scarcely conscious.

The real issue here is not where I get my power but that a life entirely disburdened is a life entirely disengaged and a life entirely disengaged is a life out of balance. I agree with Borgmann. We need disburdenment in our lives but we also need engagement. He suggests a variety of engaging practices, which he calls focal practices, like running or "the culture of the table," which though outside the immediate purview of the engineering world do depend at some stage of their coming into being on engineering processes. My suggestion is that we devote more energy directly to the spirit of the focal engineering venture as it seeks to bring into the world products that have an engaging rather than disengaging aspect. *Engagement* is a measure of harmony between end user and focal product. But there are other values important to focal products: *enlivenment,* a measure of harmony between end-user and the world, and *resonance,* a measure of harmony between the world and the focal product. Later in the book I will invoke all three values in an ethical assessment of the focally engineered product.

Consider the Golden Gate Bridge in San Francisco. It is one of the most popular and spectacular bridges in the world and the symbol of one of the most cherished of American cities. It has a commanding presence that has inspired poems, paintings, and photographs. The Golden Gate is a structure that gathers people to it, enlivens and elevates their spirit, and resonates with its setting. Humans, bridge, and world are all in tune.

The Golden Gate Bridge is an engineered structure I would call *focal.* It is engaging, invigorating, and harmonious. We shape our worlds around focal objects by virtue of our nurturing involvements with them. Focal objects are never "things as such" but are always embedded in social, cultural, or natural contexts. Focal practices, like tending to a wood-burning stove, require a regular engagement with orienting activities that demand skill and effort.

Unlike the Golden Gate and the wood-burning stove, however, most engineered systems, devices, structures, networks, and organisms are non-focal, employed for their functionality and not much else. But, in fact, that is often sufficient. Disburdenments, of course, are welcomed. I want to heat up a cup of coffee. I put it in the microwave for thirty seconds and in the meantime boot up my computer to check my email. Before the computer comes on, the coffee is done. Mission accomplished. My days are packed with these functional events. The problem, again, is not these activities per se, but that when these events constitute the whole of my being, life becomes a humdrum, leveled affair. So I seek out diversion and distraction, which result, in general, in further leveling. A focal practice, on the other hand, can enliven life, make it meaningful, bring it into relief. I write a poem, and my world lights up. If I use my computer to write that poem, I am using technology in a functional manner but also in a manner that contributes to focal reality. A computer can be focal or nonfocal, though when I speak of focal reality, again, I must think, not computer, not "thing as such," but computer/world conjunct. It is not just a matter of applying the device to a focal or nonfocal event. Patterns are what matter. Although an engineered product is seldom itself a focal object, *if* it can serve a focal practice, and be immersed in a world patterned by such practices, as exhibited in a pattern of life that is resonant, enlivening, and engaging, *then* the engineering involved can be considered focal.

Focal engineering aims to bring focal products into the world. Consider the windmills tastefully placed along the canals in the Netherlands. They require focal engagements and are focal things contextually embedded. Such products, instead of just aiming to do no harm, actually seek to contribute to the good life in a convivial society, and once the notion of the good becomes prominent, ethics and morality enter into our deliberations. The ethical assessments of focal engineering must be made from within an inclusive and hopefully democratic decision space. The corporate capitalist, fulfilling real needs or artificially manufactured ones, cannot be the only voice. Politicians, concerned citizens, environmentalists, philosophers, scientists, spiritual leaders, and others must be invited to the conversation about what constitutes a focally engineered device.

David Billington categorizes the types of entities that can be taken as engineered entities as machines, structures, networks, and processes.[4] Those distinctions can be further nuanced to cover a wide range of what Carl Mitcham calls "types of technology as object," including utensils, apparatus, utilities, tools, and automata.[5] For my purposes, these will all be in the category of *the engineered,* the products of the engineering process.

4. David Billington, *The Innovators: The Engineering Pioneers Who Made America Modern* (New York: John Wiley & Sons, 1996), 201.

5. Carl Mitcham, *Thinking Through Technology: The Path Between Engineering and Philosophy* (Chicago: University of Chicago Press, 1994), 162.

Of course, we use tools and equipment of all sorts when we do engineering. I will be primarily concerned with the engineered as a product of the process of engineering, recognizing that many of these products, like oscilloscopes and computers, serve the engineering process itself.

We engineers most commonly gauge the viability of the engineered by a technical assessment in an ongoing and immediate fashion: it is what we do when we do engineering. Efficiency, productivity, objectivity, and precision are quintessential technical measures, even though different branches of engineering might interpret these measures differently. For instance, in looking at efficiency, mechanical and electrical engineering speak of it as the mathematical ratio of output over input energy. Civil engineering views efficiency broadly in notions of accuracy of measurement and durability of structures, as well as measures of economy of cost and elegance of form. But the technical assessment looks at the rational good of the engineered in a local and limited sense. From a broader perspective, which encompasses the engineer, engineering, and the engineered, what is locally benign may present serious moral problems in the realms of social justice, environmental sustainability, and health and safety of affected parties. The moral assessments of engineers, of engineering, and of engineered products require a step back and away from immediate concerns to reflect on the larger picture of the contextually embedded engineering project. It is here where Mitcham's notion of technology as *volition* comes into play. As Mitcham maintains, technology is certainly an *activity* and employs all kinds of *objects* and involves various modes of *knowing*.[6] But, as volition, technology most clearly becomes a phenomenon we humans can take control of, fit to our purposes, and mold into good products, giving engineering the opportunity to authentically make good on the promise of technology.

Both technical and ethical assessments, then, are intrinsic to the engineering project. Both are needed in the assessment, for example, of focally engineered products. Furthermore, any particular assessment is seen most clearly in the light of a general framework, suggesting that we need to look at engineering in general and attempt to see it in the fullness of its being. Such an effort, at a minimum, entails surveying the engineering project from social, philosophical, and historical contexts. Social and philosophical contexts provide a vertical, synchronic reach to engineering, while

6. Ibid., 161–266.

the historical context provides a horizontal, diachronic range. The social and philosophical perspectives on the engineering project uncover an expanded need for ethics in engineering, a need to ask not only the conventional questions about how to *be* good engineers and how to *do* good engineering, but also, and especially, to ask about how to *make* good products. Being, doing, and making are all gathered and harmonized in the practice of focal engineering. In addition, social and philosophical perspectives reveal a structure and function that have an inward-looking *micro* aspect and an outward-looking *macro* aspect. Generally, the inward look, following Mitcham again, is the way engineers view engineering, and the outward look is the way social scientists view engineering. Structural and functional portrayals, revealing a schematic look at the engineering endeavor, will be fleshed out by the integration of historical perspectives. Those historical perspectives will point to modulations over time of the macro and micro aspects of the engineering project, as well as to perturbations within the compass of engineering ethics. To champion focal engineering, I will need to show the centrality of focal engineering in the process of engineering a world oriented toward a resonant life of engagement and enlivenment. This book will be a task of analysis as well as synthesis, and I will provide, as product of my reflections, both descriptions of and prescriptions for the project of engineering.

Since ethics plays a key role throughout the book, several chapters will be dedicated to discussions about *virtue ethics, process ethics,* and *material ethics.*[7] Each of these kinds of ethics is keyed to one of the dimensions of the engineering project: the *engineer, engineering,* and the *engineered.* Virtue ethics concerns itself with the character of the engineer, process ethics with the character of the engineering process, and material ethics with the character of the product brought into being. Though these elements cannot be separated, they can and will be distinguished. The inward, micro look at the engineering project reveals a confluence of these three dimensions.

Each type of ethics will deal with one of the issues: virtue ethics with the ambivalence often found in the hearts and minds of practicing engineers, process ethics with the ambiguity inherent in the actual practice of

7. The idea of material ethics comes from the work of Albert Borgmann. These two essays are especially useful: "The Moral Assessment of Technology," in *Democracy in a Technological Society*, ed. Langdon Winner (Dordrecht: Kluwer Academic Publishers, 1992), and "The Moral Significance of Material Culture," *Inquiry* 35 (1992).

engineering, and material ethics with dilemmas that arise in the quest to engineer focal products. Who performs the assessments necessitated by these three types of ethics? And where? I will point to three levels of assessment, including the *group or team* level, the *professional* level, and the *social* level. At the group level, engineers can do a self-assessment in terms of certain virtues that will be seen to be important, and the members of the group can also be assessed by the group leader, as is common in engineering practice today. Or instead of the group leader, a company may engage the services of an ombudsperson. At the professional level, professional societies can assess the ethicality of the engineering process and of the engineer as well. At the social level, an overall assessment of the engineering project needs to be made. This includes not only the material ethics assessment of the engineered product, but also the ethics of the process and the assessment of the engineer. Ultimately, at the social level, all three types of assessments will be involved. At this level assessment teams need to be formed that will include a broad diversity of voices.

The outward-looking, macro aspect of engineering reveals it to be a contextualized project with several nested layers of context. The three most important are the realm of *technological systems,* the realm of *systems,* and the *human lifeworld,* as indicated in Figure 1.

The engineering project is a particular type of technological system, embedded in the context of technological systems in general. There are technological systems that are not included within the domain of the engineering project. Technological systems are particular types of systems, embedded in the context of systems in general. Furthermore, the systems way of being is a particular way of being embedded in the human *lifeworld,* the realm of our everyday affairs, wherein we take up with all sort of things, and pursue goals and enact roles of every kind imaginable.[8] There are systematic and nonsystematic ways of being in the lifeworld. I can grow row after row of genetically modified corn in a seriously systematic project, or I can just sit in my backyard and shoot the breeze with my friends in casual communion. Although the notion of the human lifeworld is uncommon in engineering, it forms the background of our everyday engineering activity and is fundamental to the philosophical concerns of this text.

8. Herbert L. Dreyfus, "The Priority of *The* World to *My* World," *Man and World* 8 (May 1975): 122.

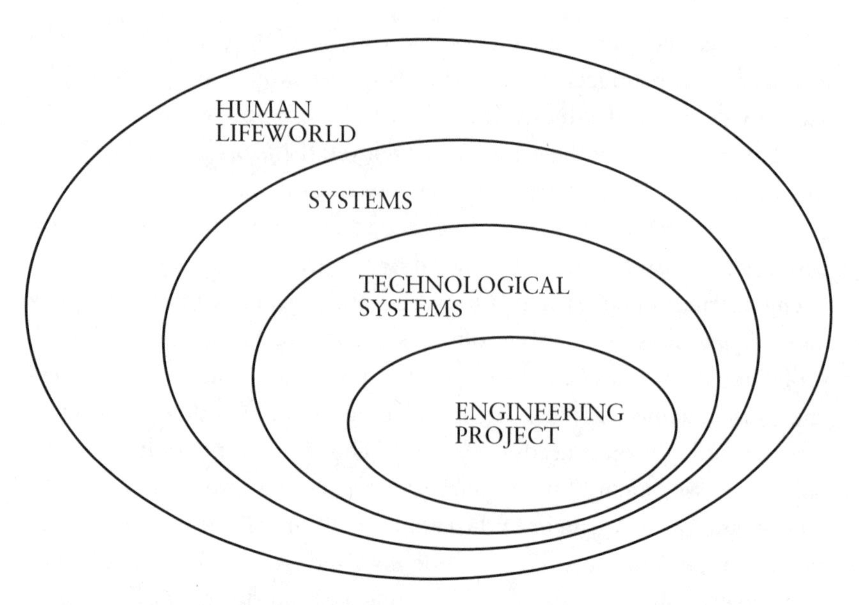

Figure 1. The Engineering Project in Context

Jürgen Habermas insists that the crucial event in contemporary times is the *colonization* of the lifeworld by the realm of systems.[9] The familiar notion of colonization refers to the process whereby one country settles in and takes over another, as Belgium did to the Congo in Africa. But Habermas refers to a more philosophical sense of the word. What he means by the notion of colonization is that as systems proliferate—many of them driven by the products of the engineering project—they impose on the fluid reality of the lifeworld, ever more stringently, the values of productivity and efficiency. Drawing on the work of Talcott Parsons, Habermas distinguishes between quantitative and qualitative media. Qualitative media involve influence and value-commitments that are only enacted in communication between people. Within such communicative action or discourse differences are hammered out and people come to more or less common understandings. Colonization diminishes the realm of qualitative and communicative discourse so that more and more of the lifeworld becomes noncommunicative and quantitative. Some of that is of course a good thing. But how much? Technology can be employed within the

9. Jürgen Habermas, *The Theory of Communicative Action,* 2 vols. (Boston: Beacon Press, 1984, 1987).

lifeworld without becoming excessively colonizing. The Appropriate Technology movement comes immediately to mind. But it is also important to keep some acreage of the lifeworld free from systematization altogether. We humans seem to need ways of being that include just hanging out, kicking back, and goofing around.

I will consider what the specific contribution of the engineering project is in the colonization process and, in light of the judgments of engineering ethics, what it ought to be. Throughout these discussions it will be important to keep alive the difference between contextualization and colonization, especially the colonization of the lifeworld by the engineering project and the contextualization of the engineering project by decisions made in the human lifeworld. A distinction Andrew Feenberg makes between what he calls the Primary and the Secondary Instrumentalizations resonates with the contextualization/colonization distinction and will be useful to my concerns.[10] The Primary Instrumentalization involves the processes of abstraction or de-worlding or decontextualization of a sector of concern from out of the human lifeworld, and deconstruction of that sector into basic and useful parts or particles. These processes are driven by ideals of rendering the lifeworld more efficient and productive. The role of the Secondary Instrumentalization is to reconstitute reality by reintegrating useful entities, which for us would be the products of the engineering project, into natural or artificial systems. Those systems need to be controlled by decisions made in the lifeworld. The quintessential example is nanotechnology, which de-worlds matter and breaks it up into atoms and molecules, which are then reassembled (ideally) into any desired object and those objects are reintegrated into wider systems and the human lifeworld. Lifeworld decisions contextualize the engineering process. System colonizes lifeworld at the same time that lifeworld contextualizes system.

The text proper has three parts, and each part has three chapters. The first part investigates the modern engineering enterprise, with emphasis on the process of engineering and the nature of process ethics, and explores the idea of the colonization of the lifeworld by systems, especially the modern engineering enterprise in alliance with the system of corporate capitalism. The second looks back in time to the premodern engineering endeavor, with emphasis on the person of the engineer and virtue ethics,

10. Andrew Feenberg, *Questioning Technology* (London: Routledge, 1999).

and explores the idea of contextualization, which was more prominent within the premodern engineering endeavor than it is now. The third looks forward to the focal engineering venture, with emphasis on the product and material ethics, and explores the possibility of striking a balance between contextualization and colonization.

There is much talk nowadays that laments an apparent decline in engineering professionalism. One hears much discussion about new graduates from engineering schools being minimally competent technically and poor in communication skills. Strategies for addressing these problems abound, and the struggle continues, as well it should. Nevertheless, I propose that professionalism must be advanced, not only by boosting the skill levels of engineering graduates, but also by instilling in them an understanding of engineering as structurally and historically rich in meaning, as ethically variegated, and as both a colonizing and contextualized project. As such, engineering is capable of an orientation toward a more exhilarating and exalted nature, toward a focal engineering whose products embellish the lives of end-users and, more generally, contribute to the good life in a convivial society, a society in which hospitality and a spirit of trust can begin to flourish.

PART I

THE MODERN ENGINEERING ENTERPRISE

ONE

PROCESS

In the premodern era, which lasted roughly from the time of the pyramids to the heyday of the medieval cathedral, engineering processes of a basic sort did exist. Engineers had a kind of knowing that can be described as know-how, which was embedded in skills of various sorts. In building their aqueducts, for instance, Roman engineers had to know how to locate sources of water. They had to know how to dig tunnels. They had to know how to construct pumps. They had to know how to purify water. They acquired this know-how as apprentices, through experience, or by trial and error.

However, toward the end of the medieval period, engineering know-how began to couple with the know-what of the modern scientific mind, which was just then beginning to bloom. The premodern engineering endeavor began its transformation into the modern engineering enterprise. Means, methods, and procedures within the process of engineering began to become explicit, and mathematical and scientific theory to reveal, in what was assumed to be a clear and distinct fashion, the essence or *whatness* of the universe. At the same time, the engineering enterprise gradually came into its own as a unique practice, thanks in large measure to the emergence of an increasingly clear and distinct methodology.

The scientific method, in broad outline, was adopted by the engineering enterprise. Of course, what constituted the scientific method originally and ideally has today been recognized as less of an objective procedure and more of a human practice involving imagination, speculation, intuition, heuristics, and often just plain luck. All of these human elements, encompassing both know-how and know-what, are intrinsic to contemporary real-world engineering practice. However, René Descartes (1596–1650) was the first to formulate a methodology that, in a rough and ready sense, became useful to engineering. Descartes's method consists of four rules: abstraction, dissection, reconstruction, and control.[1] The first two are fundamental to the practice of *analysis,* the second two to *synthesis.* Through analysis, science enters the program of engineering in an explicit manner. The method instructs the practitioner to abstract a representation of a realm of concern from its context or world, and then dissect that abstracted representation into its fundamental parts. These practices are for the sake, initially, of scientific understanding. As the engineers understood it, however, this scientific understanding was inevitably for the sake of subsequent engineering developments. For example, in contemporary engineering practice, analysis of a circuit yields an understanding of the power and energy associated with various circuit elements. But that understanding is in the service of developing, say, a control system that uses the circuit but has another larger task assigned to it, like stabilizing an aircraft. Although the knowing-how of the premodern engineer was still important, it was scientific knowing-what that came to characterize the analytic side of modern engineering.

The synthetic aspects of the modern engineering process are derived from the rules of reconstruction and control of the Cartesian method. When engineers reconstruct, they assemble the parts separated in analysis. The goal is to synthesize a system, in a timely manner, that meets the specifications of a customer. Even though the know-how of premodern engineering synthesis is now coupled with the know-what of the modern engineering enterprise, this does not mean that engineering has become a science or even an applied science. The fundamental knowing-how, which had always been at the heart of the engineering project, was refined, not replaced, by scientific knowing-what. Modern engineering employs what might be called a scientifically informed heuristic procedure applied to a collection of activities, including design, testing, production, manufacturing, marketing, maintenance, and control. All these activities are at first contextually situated and constrained, but once the method swings into gear, it seeks to elude its contextual constraints. If the methods of the modern engineering enterprise are truly de-worlded, then their unfoldings proceed in a straightforward and unencumbered fashion. That is the way, for the most part, that engineering is taught in engineering schools.

1. Albert Borgmann, *Crossing the Postmodern Divide* (Chicago: University of Chicago Press, 1992), 35.

Although it embraces scientific methods, engineering itself is not a science. Unlike engineering, science seeks objective and ideally value-free truth for its own sake. Science operates at the level of theory, a theory of reality. A science of engineering, according to Taft Broome,[2] would have to be a praxiology, a theory of efficient action. But how useful would such a science of engineering be? Engineering as the practice of making good on the promise of technology is an action involving a process. That process is a volitional human activity of producing a plan, which draws on resources available and involves a variety of ways of knowing how to produce systems, devices, networks, organisms, and structures to fulfill human needs and desires. A theory of efficient action could encompass only the barest outlines of such a complex reality because modern engineering is, as Steven Goldman puts it, "shaped by manifestly arbitrary, that is, explicitly willfully imposed constraints expressive of a host of personal, institutional, social (including economic and political), and cultural (including aesthetic and religious) value judgments."[3] Thus, in this book I will try to describe not a pure science of engineering but the possibilities that unfold in processes of the engineering project as it is actually practiced.

A science of engineering is one thing. Engineering science is quite another. Engineering science is science drafted into the service of engineering processes. For example, engineering thermodynamics is very different than thermodynamics taught in a physics department. The physics version is highly theoretical. The engineering version includes engineering theories, descriptive regularities, engineering laws, and maxims.[4] The physicist studies thermodynamics for its own sake. The engineer studies it to further the process of engineering.

Engineering science is only one part of the manifold of disciplines and activities that contribute to an engineering process. Of all the subprocesses associated with engineering, like testing, design, prototyping, production, manufacturing, marketing, maintenance, and so on, design is commonly considered the essence or the heart of the entire matter. Design is one of my focal points in this chapter. Goldman maintains that design

2. Taft H. Broome, "Engineering the Philosophy of Science," *Metaphilosophy* 16, no. 1 (January 1985): 48–49.

3. Steven L. Goldman, "Philosophy, Engineering, and Western Culture," in *Broad and Narrow Interpretations of Philosophy of Technology*, ed. Paul T. Durbin (Amsterdam: Kluwer Academic Publishers, 1990), 131.

4. See Mitcham, *Thinking Through Technology*, 192–208.

"is unquestionably central to engineering, and design is an explicitly valuational activity, a necessarily non-unique synthesis of a 'box' of means, a set of imposed constraints, some natural, most arbitrary, and a fuzzy vision of an end to be achieved."[5] Closely tied to the design process, production engineering, which brings the designed artifacts of the engineering enterprise into the world, is another essential part of the engineering process: there is not only increasing concern for production of designs but also for design of the production process itself.

The question arising throughout the development of the modern engineering enterprise is, How *ought* the process to proceed? How is the engineering process to be gauged or assessed? How positive or negative, how right or how wrong might the process be? The process could be technically precise and efficacious and at the same time ethically problematic. The incorporation of ethical standards within engineering went hand in hand with its professionalization. Engineers became less and less craftspersons, artisans, or foremen overseeing engineering projects, as they had been in premodern times, and more and more professionals implementing engineering processes. The establishment of professional engineering societies in the nineteenth century, including the American Society of Civil Engineers (1852), the American Society of Mechanical Engineers (1880), and the American Institute of Electrical Engineers (1884), established a forum for the resolution of ethical issues. The ethics of the day sprung from the work of Jeremy Bentham (1748–1832) and John Stuart Mill (1806–1873), who developed and promoted the philosophy of utilitarianism, which advises us to do whatever advances the greatest good for the greatest number. Also, a generation before Bentham and Mill, Immanuel Kant (1724–1804) gave ethics the concept of the Categorical Imperative, one version of which advises us to act in such a way that, if everyone did the same, the Good would be served. These notions of ethics, which I call process ethics, provided moral guidance for engineers carrying out the processes of the modern engineering enterprise.

Process ethics came into the service of modern engineering quite naturally because modern engineering and process ethics were both grounded in the general scientific and theoretical mindset that characterized the early modern worldview. The emerging professional engineering societies struggled to balance their freedoms and their responsibilities, and these struggles

5. See Goldman, "Philosophy, Engineering, and Western Culture," 131–32.

crystallized into codes of engineering ethics. Today, the dimensions of process ethics are encapsulated succinctly in a number of professional engineering codes, such as the IEEE Code of Engineering Ethics.

SETTING THE STAGE

The cultures of the ancient world produced vigorous technology, which resulted in an abundant material lifeworld. But as resources, especially forestry resources, began to be depleted, and populations grew enormously, especially due to influxes of invaders from the north, ancient cultures went into decline. However, the real cause of that decline, according to Taichi Sakaiya, whether we are talking about the Western Roman empire or the Western Jin dynasty or ancient India, was ultimately not the invasions but "the change in tastes and ethics that had been under way long before that."[6] In Western Europe, antiquity was superseded by the Middle Ages, a period in which God, not man (or woman), was the measure of all things. The medieval period lasted until roughly the mid-fourteenth century and was followed by the Renaissance, which exalted the human spirit mainly through art and literature. The modern era, as least as far as engineering is concerned, began between the beginning of the sixteenth century when da Vinci was working and the end of the sixteenth century when Galileo set experimental physics on its true path.

The transition from antiquity to the medieval era was marked by a movement away from materialism toward a more spiritual orientation. The medievals in looking backward saw a progressive advance of values developing into a better culture, which integrated classical cultures with a faith in the Christian God. Borgmann, in fact, maintains that Charlemagne in the ninth century had consolidated the medieval epoch in Europe by fusing together classical culture and learning, the feudal order of Germanic tradition, and a vision of Christianity.[7] The classical cultural content of the medieval era was tremendously deepened as the Renaissance began to spread over the European continent a few centuries after Charlemagne. Yet in spite of this affinity with classical ideas, which had the pretense of a logical and objective worldview, most medieval societies

6. Taichi Sakaiya, *The Knowledge-Value Revolution,* trans. George Fields and William Marsh (New York: Kodansha International, 1991), 162.

7. See Borgmann, *Crossing the Postmodern Divide,* 21.

emphasized a kind of social subjectivity. Fantasy and dream life and what we today might call an overactive imagination tended to prevail over rational and empirical explorations of reality. With medieval investigators, Sakaiya maintains, very seldom were there "even attempts to comprehend things in terms of the sight of one's eyes or the touch of one's hands upon the things of this world. Their explorations belong to another realm entirely."[8]

Furthermore, the orientation toward subjectivity in medieval cultures, even though it was predominant in Europe, can also be found in India, China, and the Islamic Middle East during this period. The realism of the ancient art of Greece, Rome, India, and China had given way to a kind of naive symbolism. Subjectively describing one's feelings was taken to be more important than materialistic objectivity or concreteness or accuracy. But the medieval period lasted only a few hundred years. When and why the medieval period ended are questions that have no unique answers. The medieval civilizations were bound by religious traditions and local values and mores, all of which were called into question as the effects of the secular humanism of the Renaissance began to be more and more widely felt.

For the most part, the lives of medieval people were narrow and simple, especially if compared to the vast possibilities of modern life. However, they did live rather vigorously, being guided by various excellences, like chivalry and courtesy, community and celebration, authority and craft.[9] But that vigor was short-lived. It had run its course by the late 1400s. As Borgmann puts it, "Unlike the slow and convulsive decay of Greek and Roman culture, the medieval form of life came to a swift and unambiguous end. It was shattered by the three blows that we commonly associate with Columbus, Copernicus, and Luther."[10] The direct force of these men, coupled with the more indirect force of the spread of secular humanism, proved to be irresistible.

With the rise of the new *Zeitgeist* of modernism informing European cultures—and Sakaiya shows similar transformations occurring, not necessarily at the same time, in China, India, and the Middle East—the fusion that Charlemagne had achieved began to be diffused. A new spirit of discovery and adventure, exhibited in the enterprising boldness of Christopher Columbus (1446–1506), gripped the European mind. A new

8. See Sakaiya, *The Knowledge = Value Revolution*, 166.
9. See Borgmann, *Crossing the Postmodern Divide*, 21.
10. Ibid.

Weltanschauung was distilled out of the discoveries of the Polish astronomer Nicolaus Copernicus (1473–1543). And Martin Luther (1483–1546) broke the grip of the Church of Rome on the northern half of Europe by denying the need for Catholic tradition and putting scriptural interpretation in the hands of the individual.

Borgmann points out that "unlike the fallen empire of Rome, the shattered Middle Ages did not lie in ruins for long. Less than a generation separates the last of the destroyers of the medieval order from the first of the founders of modernity, Francis Bacon."[11] Leonardo da Vinci might also be seen as a founder of modernity. Da Vinci died in 1519 and Bacon was not even born until 1561. Even though da Vinci is typically associated with the late Renaissance, there was some overlap between the Renaissance and modernism. In any event, among others, Bacon, da Vinci, Descartes, Locke, and Galileo certainly should be considered pivotal contributors to the beginnings of the project of modernism.

MODERNISM

Toward the beginning of the sixteenth century several more or less simultaneous transformations in thoughts, words, and deeds brought about the modern era. Modernism resulted from an array of new discoveries in the natural sciences which shifted our image of ourselves and of our place in the universe. Modernism was also generated by the rising spirit of productivity and the industrialization of production, transforming scientific knowledge into technology (which is where modern engineering came into the picture), creating new human ways of being and environments and destroying old ones. With the rise of modernism there was an acceleration of the whole pace of life. New forms of corporate power and class struggle emerged. As Marshall Berman puts it, there were

> immense demographic upheavals, severing millions of people from their ancestral habitats, hurtling them half-way across the world into new lives; rapid and often cataclysmic urban growth; systems of mass communication, dynamic in their development, enveloping and binding together the most diverse people and

11. Ibid., 22.

> societies; increasingly powerful national states, bureaucratically structured and operated, constantly striving to expand their powers; mass social movements of people, and peoples, challenging their political and economic rulers, striving to gain some control over their lives; finally, bearing and driving all these people and institutions along an ever-expanding, drastically fluctuating capitalist world market.[12]

But how does engineering fit into all this? Modern engineering, seeking ever more aggressively to make good on the promise of technology, augmented the know-how typical of the premodern engineering endeavor, with the know-what of science. The theory and practice, or *theoria* and *praxis,* distinction was brought into sharp relief within the nexus of forces constituting modernism. Initially, it was practical mathematics that broke with the purely abstract Platonic kind of mathematics, generating a more grounded *theoria.* Also, there arose a new array of measuring instruments that began to aid the observational scientist. Clocks became popular and led to a new precision not only in mechanisms but also in the control and planning of human activities in everyday life. In the seventeenth century the telescope, the thermometer, the vacuum pump, and the microscope all came into prominence.[13] Such sophisticated instruments led to a new kind of *praxis,* one based on sound technology. *Theoria* and *praxis,* although they were sharply distinguished in the modern era, actually became more intimate and intertwined, especially compared to the very different worlds they had inhabited in ancient and medieval times. In the premodern era, scientists or natural philosophers had seen little relationship between their work and the practical business of making or "engineering" artifacts. But with the dawning of modernism, a wedding was called for that would unite the purely scientific idea of describing the world, ideally in a language of mathematics, and the practical processes of shaping the human environment.

The application of the new science, then, was not only liberating but also promised to procure the good life. Of course, it was not science, per se, but rather technology, by virtue of its manipulation of external realities,

12. Marshall Berman, *All That Is Solid Melts into Air* (New York: Simon and Schuster, 1982), 16.

13. John Lienhard, *The Engines of Our Ingenuity* (Oxford: Oxford University Press, 2000), 71.

which drove that procurement. The new technology, aided by its alliance with the new science, began to develop into the diverse fields of modern engineering. Science and modern engineering do not interact so much in a causal sense, but more in the sense that they share a common ground: they both derive largely from the human will to master and exploit nature for the human ends of gaining knowledge or procuring the good life.

Engineering activity profited from science and mathematics initially as an aid to the practices of planning and building. One of the first disciplines to embrace the marriage of theory and practice was architecture. When mathematics was joined to masonry toward the end of the fifteenth century, according to John Lienhard, "a new baroque architecture emerged—one based on exact geometrical methods. Then architects started using precise intellectual apparatus to design magically spatial forms: barrel vaults, biased arches, helicoids, and embellished versions of the medieval trumpet squinch."[14]

Later in the modern era, the incorporation of mathematics and science into engineering praxis served the analysis and design processes that were becoming crucial to the engineering enterprise. Ultimately, the incorporation of mathematics and science into engineering was seen as essential to all authentic engineering practice. That incorporation became what distinguished the engineer from the technician. The disciplines of mechanics, thermodynamics, materials science, electromechanics, and especially today, computer and information sciences, all serve modern engineering. Mathematics, queen of the sciences, in her applied aspect, also serves engineering. Today these sciences in the service of engineering are called *engineering sciences*. Many were the contributors to the engineering sciences and to the modern engineering project in general. One of the first was Leonardo da Vinci.

MODERNS

Leonardo da Vinci (1452–1519) brought the scientific rigor of mathematics to bear on his diverse artistic and practical endeavors. The prototype of the Renaissance man, the scope of his endeavors was enormous. He made contributions in several fields, including sculpture, painting,

14. Ibid., 71–72.

graphic design, aviation, and machine design. His notebooks are famous for their exquisite illustrations not only in the realm of the fine arts but also in the realm of technical structures. As historian Friedrich Klemm points out, da Vinci's "familiarity, thanks to the Florentine workshop tradition, with the properties of different substances and the varied possibilities of their utilization in the workshop, and his labours to discover by experiment simple mathematical laws of nature, all these made him an engineer in the modern sense."[15]

Leonardo suggested that the universe was rational and measurable. Although most of his works, especially his art, consisted of things that embellished specific local settings, as was the case with medieval art in general, as one of the first modern engineers, Leonardo brought mathematical space and time considerations, tending toward universal applicability, to bear on his projects more intensely and consistently than had any engineer before him. Still, it remained unclear to da Vinci and his contemporaries exactly how mathematics was to be applied to nature and why this application should yield a superior type of knowledge. The work of Francis Bacon helped to clarify these issues.

Francis Bacon (1561–1626) articulated what da Vinci hinted at in his works, namely, the importance of the scientific attitude, especially the utility of using science to procure knowledge and the application of that knowledge to the practical problems of everyday human life. Bacon believed that human misery was not the will of God but an unnecessary indignity that science, technology, and rationality could defeat through the subjugation of nature.

Although Bacon aggressively sought to put theory in the service of practical human life, theory could not marry practice without each first refining the other. Before Bacon's time, the ways of practice were intuited, for the most part. They were beyond the reach of explicit conceptualization and were based largely on experience. In the utopia Bacon proposes in his treatise *New Atlantis* (1627), mere experience has been replaced by rigorous experiment and conceptualization. Before Bacon's time, pure theory was itself often frivolous, a speculative and unfounded pondering of imponderables, like endless speculations concerning the distinction between material and immaterial reality. Bacon advocated clearing the

15. Friedrich Klemm, *A History of Western Technology*, trans. D. W. Singer (Cambridge: MIT Press, 1964), 124–25.

speculative ground to establish a *tabula rasa* on which the arts and sciences could be reconstituted. W. T. Jones, in his *History of Philosophy,* quotes Bacon as saying that a clean slate would be obtained by purging knowledge "of two sorts of rovers, whereof the one with frivolous disputations, confutations, and verbosities, the other with blind experiments and auricular traditions and impostures, hath committed so many spoils, I hope I should bring in industrious observations, grounded conclusions, and profitable inventions and discoveries."[16]

Industriousness in applying science to the advancement of the human condition was Bacon's central contribution to the modern engineering project. But even "the patron saint" of the modern engineering enterprise put forth ideas that were by no means without controversy. His own proposals, especially his highly touted inductive method, left much to be desired, because in his proposals he assumed a notion of *substance,* precisely one of those baseless holdovers from the medieval era that he had condemned. Substance is a rather vague term but can be thought to mean "essence." In other words, Bacon set out to study natures or essences, and developed a method that revealed *relations.*[17] It would take Descartes to provide not only a clear understanding of substance—in terms of extension and motion of any body at issue—but also the method proper for the modern engineering project.

René Descartes not only expressed optimism about science and its power to describe and explain the world, but also firmly believed that science would improve the human condition. Descartes and Bacon were in accord on this. Bacon said what we should do: subjugate nature to advance the human estate. Descartes showed us how to do it: employ a rational method in an orderly manner based on clearly stated and distinct ideas. Bacon was still caught up in the particulars of his lifeworld. Descartes's methodology was universal. Bacon's inductive method began and ended with concrete particulars. Descartes's method of abstraction was, as Borgmann puts it, "the triumph of procedure over substance."[18]

The Cartesian method, which privileges science and the scientific worldview, was exactly what the nascent modern engineering needed in order

16. W. T. Jones, *A History of Western Philosophy: Hobbes to Hume,* 2nd edition (New York: Harcourt, Brace & World, 1969), 74.

17. Ibid., 86–87.

18. See Borgmann, *Crossing the Postmodern Divide,* 24.

to give itself legitimacy. The war against nature was already in full swing with Bacon leading the charge. Borgmann again: "The spirit of domination was pervasive, yet its works were various and scattered. As the conquests grew and were consolidated, there was a need for large-scale integration. Baconian aggressiveness began to require the complement of Cartesian order."[19] The Cartesian method, as it began to be applied within the engineering project, helped to shift engineering from being a tradition-bound endeavor involving a dispersed set of skills and general rules to being a scientifically grounded enterprise, on its way to being a respected and honorable profession.

Galileo Galilei (1564–1642) was another key player in the early development of the modern engineering project. Though known mostly as a physicist and astronomer, Galileo was also interested in practical problems. As founder of the new dynamics, his theoretical work stemmed from real technical problems. Galileo pursued studies in physics exhibiting a combination of experimental research and technical applications.[20] In Galileo's time, physical science and philosophy were closely tied, and philosophy of the speculative and Aristotelian order provided the starting place for investigations into the physical world. The problem was that much speculation was off the mark and a lot of Aristotle's presuppositions were flawed. Galileo is credited with providing the definitive separation of physical science from philosophy. In fact, his writings mark the beginning of a movement that opposed speculative, occult, and metaphysical philosophy. Instead of continuing to call themselves natural philosophers, investigators eventually started to refer to themselves as scientists. And these early scientists had much affinity with the early modern engineers.

The "new" science was far from what we would call a pure science. Galileo, after securing an academic position at the university in Padua where he taught from 1592, occupied himself with a diverse set of technical problems such as fortifications, the technique of water supply, the mechanism of simple machines, widely usable proportional compasses, and the testing of materials.[21] These highly practical affairs did not involve just know-how in the traditional engineering sense, they were also served by the know-what of the new scientific concept that physical reality

19. Ibid., 34–35.
20. See Klemm, *History of Western Technology,* 175.
21. Ibid.

consists of matter in motion. The dimensions of matter, its velocity, and its weight, according to W. T. Jones, "were what were measured in physics, and the result of these measurements showed that its behavior conformed to simple mathematical 'laws.' For the qualitative and teleological conception of nature with which men of the Middle Ages had operated, there was thus substituted a quantitative and mechanical one."[22] These advances in the scientific concept of reality, to which Galileo was a major contributor, laid the groundwork for the development of the engineering sciences, which were fundamental to the process of the modern engineering enterprise.

John Locke (1632–1704) was another major modern personage whose work impacted the program of modern engineering. When at the dawn of modernism both church and state began to loose their authority and came into question as providers of the foundation of the common order, Locke stepped in to fill the breach by proposing the sovereignty of the individual as the ultimate source of authority. Unlike the Baconian and Cartesian bequests to modernity, which were quite straightforward, Lockean individualism has always been a complicated affair.[23] What makes individualism so complicated is the problem of sociality: how to get the one to be part of the many. Locke had many solutions to this dilemma, but none was very satisfying and the concept of individualism has remained to this day fraught with ambiguity. A more or less common order does exist but, according to Locke, only through an agreement or contract of individuals. The individual always had the last word, the individual as both producer and as consumer. The alternative view, more popular in premodern cultures, was that the collective was primary. Individuality entailed a separation from the whole and was even sometimes seen as deviant.

In any event, it is precisely the individual as producer, aggressively challenging nature to yield her resources, who pushes forward the modern project.[24] The concept of individualism, championed in theory in clear and distinct terms, though problematic in practice, plays a major role in the project of modern engineering, as it does in all modern discourse.

22. See Jones, *History of Western Philosophy*, 115.
23. See Borgmann, *Crossing the Postmodern Divide*, 37–38.
24. Ibid., 38.

THE MODERN ENGINEER

Though the process of modern engineering is my focus in this chapter, I want to say a few words about modern engineers. They seek to develop devices, structures, and systems that will advance the human estate. To do so, they rely on the aggressive subjugation of nature in order to procure the needed resources. They employ a universal methodology—in a pragmatic manner—involving abstraction, dissection, reconstruction, and control. In the process, modern engineers are trained essentially as individuals. Nevertheless, there is a growing emphasis on team and group work in engineering education, indicating a shift from individual separateness to a communal connectivity. But in modern engineering practice, the team is more an aggregation of individuals temporarily united by a common purpose, than a community inextricably united by a common being. The focal engineer, the engineer of the future, as I will show, seeks to work more on the latter type team.

In order to produce engineers who function effectively within the culture of modern engineering, engineering education aims to inculcate in potential engineering students a set of core principles such as those advanced by Barry Lichter:

1. A concern for the efficiency of practical means;
2. A commitment to concrete problem-solving, constrained to some necessary degree by time and available resources;
3. The pursuit of optimal technological solutions based on scientific principles and/or tested technical norms and standards;
4. The pursuit of creative and innovative designs; and
5. The development of new tools for the accomplishment of each of these.[25]

Most of these principles stem from application of Cartesian methodology. Such a core presupposes the domination of nature by individuals, and the principles involved are ultimately aimed at devices, systems, and structures that serve the individual as consumer.

However, such a morally neutral core fails to address the ethical obligations and responsibilities that make engineering a profession. These

25. Barry Lichter, "Safety and the Culture of Engineering," in *Ethics and Risk Management in Engineering*, ed. A. Flores (Lanham, Md.: University Press of America, 1989), 214.

obligations and responsibilities are typically incorporated in codes of ethics. Among other things, such codes exhort engineers to heed the health, safety, and welfare of the general public. But, as Lichter maintains, "such codes have a low visibility in the engineering community, and there is little evidence that they significantly affect either engineering practice or education."[26]

Now I will look further at the process of the modern engineering enterprise. I will concern myself with these essential features: engineering *science,* which serves the process of engineering; engineering *design,* which exemplifies the process; and engineering *professionalism,* which distinguishes the process.

ENGINEERING SCIENCE

Engineering science, which includes some natural science and mathematics, as well as disciplines like circuit analysis based on fundamental scientific principles, serves the process of engineering in manifold ways. Practicing engineers use mathematics almost everywhere, for example, in sizing up maximum loads for trusses, in calculating liquid flow into pipes, in figuring out the power delivered to a load. What is rather amazing is the fact that, in spite of the complexity essential to many of the engineering sciences, the mathematics usually reduces to basic algebra. The difficulty lies in setting up the problem and applying the proper principles and procedures. For example, in analyzing an electrical circuit, to arrive at the point of having only a simple algebra problem to solve, you must understand Ohm's law, know what the passive sign convention is, write proper nodal or mesh equations, and simplify them. Then, in some cases, you need to transform the resulting equations via a method like Laplace Transformation. If and only if all these steps have been completed properly, will you arrive at a simple algebra problem to solve.

At the core of engineering science is scientific research in the field of engineering with the aim of inventing or discovering new ideas or laws that will add to the storehouse of engineering science knowledge. When doing engineering research, there is, or ought to be, an eye out for possible applications to actual engineering practice, typically design. The design

26. Ibid., 215.

problem, in fact, often provides the motivation and direction for the pursuit of engineering research. Imagine a team of controls engineers assigned the task of designing a control device employing a neural-net self-tuning algorithm. They research the literature and find the latest procedures, then they adapt one to their needs. In the process, they invent a way to perform a function in half the time it takes in the state of the art methods currently available in the literature.

Engineering research, then, is never conducted in pursuit of knowledge for its own sake, but always in the service of actual or possible engineering practice. Theory and practice as separate ventures, of course, have no place in the modern engineering enterprise. David F. Channell's essay about W.J.M. Rankine provides an interesting illustration of this fact. Rankine, a champion of harmonizing theory and practice, was one of the leading figures in the creation of the engineering sciences. He was a professor at the University of Glasgow between 1855 and 1872. During the 1860s the engineering community was split over the question of when to cut off the supply of steam in a steam engine. One approach was to apply scientific laws directly to the problem. The other was to employ traditional technology and rules of thumb. Rankine "argued for the creation of an intermediate mode of knowledge, to be developed within the University."[27]

By merging theory and practice, Rankine moved forward the development of the engineering sciences of thermodynamics and applied mechanics. Rankine recognized that both the properties of steam and the creation and disappearance of a state of heat in that steam were crucial phenomena in the workings of a steam engine. In such a system the laws of heat relied on formal theoretical concepts and an understanding of the properties of steam relied on practical experimental data. By marrying theory and practice, Rankine was actually enacting the program proposed by Francis Bacon three centuries earlier.

Engineering sciences working within the theory/practice marriage incorporate various ways of knowing. Mitcham has distinguished four separate types of technology-as-knowledge, which are applicable to engineering science. The engineering scientist embodies these ways of knowing, which Mitcham calls (1) sensorimotor skills; (2) technical maxims, rules of thumb, or recipes; (3) descriptive laws or technological rules (which for

27. David F. Channell, "The Harmony of Theory and Practice: The Engineering Science of W.J.M. Rankine," *Technology and Culture* 23, no. 1 (January 1982): 45.

engineering typically are descriptive regularities or engineering laws of the form: if A, then B); and (4) technological theories.[28]

To do engineering science is to employ some or all of these ways of knowing, implicitly or explicitly focused on and adapted to particular problems. Sensorimotor skills or *technemes,* as they are sometimes called, often involve unconscious awareness, or feelings, which may not even be considered knowing or a part of engineering science. Yet, according to Michael Polanyi, this "tacit dimension" is an integral part of all forms of knowing.[29] This subsidiary tacit dimension includes a number of sensorimotor skills involved in making and using. These skills, acquired by intuition, trial and error learning, or apprenticeships of various sorts, constitute a preconscious knowing-how as opposed to a more cognitive knowing-that or knowing-what. For example, when using an oscilloscope, an engineer has a tacit sensorimotor awareness of the intricacies of the machine while focusing on what it is being used for: observation, say, of the relative time delay exhibited by the voltages across two circuit elements connected in series. In tacit knowing, which lies more toward the feeling end of the thinking/feeling spectrum, we attend *from* elements of which we are subsidiarily aware in order to attend *to* something of which we are focally aware. Throughout his work Polanyi has elaborated the structure of tacit knowing in considerable detail.[30]

Tacit knowing is operative in an implicit way in the other forms of knowing within engineering science: maxims, laws, rules, and theorems. An example of an engineering theorem is Thevenin's Theorem from the discipline of circuit theory in electrical engineering. Thevenin's Theorem makes it possible to reduce any electrical circuit, regardless of its complexity, to a single voltage source in series with a single impedance. The theorem itself relies on various voltage and current laws and the physical law of conservation of energy.

An example of descriptive regularities or engineering laws of the "if A, then B" form is found in the part of Artificial Intelligence known as expert systems. An expert system, which is a computer program using

28. See Mitcham, *Thinking Through Technology,* 193.

29. Michael Polanyi, *The Tacit Dimension* (New York: Doubleday, 1966).

30. See not only Polanyi, *Tacit Dimension,* but also his magnum opus, *Personal Knowledge: Toward a Post-Critical Philosophy* (Chicago: University of Chicago Press, 1958). A good overview is presented in Harry Prosch, *Michael Polanyi: A Critical Exposition* (Albany: State University of New York Press, 1986).

conceptually represented domain-specific knowledge to solve problems with a competence close to that of a human expert, consists of a number of these kinds of laws. Knowledge acquisition for expert system development, involving the representation of a large body of knowledge in a precise set of rules usually of the "if A, then B" form, is the task of the knowledge engineer.[31] These gathered rules are considered empirical laws and are usually arrived at through extensive interviewing and observation of an expert in a particular field of specialization. The knowledge engineer then orders the laws into an algorithm that can perform the same operations that the expert performs.

As an example of an engineering maxim or rule of thumb, consider a very old maxim from the eighteenth-century British engineer John Smeaton quoted in a paper by Edwin T. Layton: "In a given undershot wheel, if the aperature whence the water flows be given, the effect is as the cube of the velocity."[32] These kinds of maxims are attempts to articulate generalizations about certain engineering phenomena. Maxims and descriptive laws are both empirical and tend to shade into each other. Maxims are closer to the concrete world from which they derive, while descriptive laws are closer to a theoretical framework, but they are not exactly scientific because the necessary conceptual framework from which the law could be derived is not yet explicit.[33]

A rather exhaustive and authoritative list of the engineering science disciplines that encompass the various ways of knowing is provided in the *Handbook of the Engineering Sciences,* edited by James H. Potter. Engineering sciences

> include what are called the basic engineering sciences (mathematics, physics, chemistry, graphics, statistics, theory of experiments, and mechanics) and the applied engineering sciences (thermal phenomena, heat and mass transfer, electrical energy conversion, turbomachinery, nuclear reactor engineering, aeronautics and

31. E. A. Feigenbaum, *Knowledge Engineering: The Applied Side of Artificial Intelligence,* Report STAN-CS-80-812 (Department of Computer Science, Stanford University), 1980. "Knowledge engineering" is an expression popularized by Feigenbaum, who was an early researcher in the field of expert systems.

32. Edwin T. Layton, "Mirror-Image Twins: The Communities of Science and Technology in 19th-Century America," *Technology and Culture* 12, no. 30 (October 1971): 566.

33. Carl Mitcham, "Types of Technology," *Research in Philosophy and Technology* 1 (1978): 256.

> astronautics, field theory, electromechanical energy conversion, physical electronics, electronic circuits, system dynamics, materials science, machine elements, control systems, operations research, information retrieval, preparation of reports, computers.)[34]

Fundamental to many of these disciplines are experimental practices and processes of forecasting and prediction. In distinguishing these in science and engineering, Mario Bunge maintains that scientific prediction is a means for confirmation of a scientific theory, while engineering forecasting, by suggesting how to influence circumstances, is a means of control.[35] As far as experiments are concerned, the scientific experiment aims at testing the truth of some theory, while the engineering experiment aims at testing for effectiveness. Of course, the notion of effectiveness of an engineering experiment can vary widely depending on context.

A sharper view of the distinction between engineering science and natural science can be exhibited by comparing how engineering and physics treat problems in thermodynamics. Thermodynamics is usually taught both in physics departments and in mechanical engineering departments. On the one hand, physics thermodynamics covers topics like enthalpy, local entropy production, isothermal processes, equilibrium in homogeneous systems, heat conduction in anisotropic media, and the Gibbs-Duhem Equation. On the other hand, engineering thermodynamics covers combustion, fuels, furnaces, compressors, steam power cycles, flow in nozzles, steam turbines, gas compressors, refrigeration, air conditioning, and internal combustion engines. Note that the physics topics are more abstract than the engineering topics and are not directly related to specific artifacts like turbines and engines.

The engineer, unlike the physicist, is tied to a number of real-world constraints. Engineers are paid by their employers to design and produce useful artifacts. The constraint of the bottom line holds whether engineers base their analyses on thermodynamics or fluid mechanics. In addition, the problems that arise often present serious conceptual difficulties that complicate the analysis and allow only overall simplified results. Though the devil may hide in the details, overall results may often be

34. James H. Potter, ed., *Handbook of the Engineering Sciences* (Princeton, N.J.: Van Nostrand, 1967).

35. Mario Bunge, "Toward a Philosophy of Technology," in *Philosophy and Technology*, ed. Carl Mitcham and Robert Mackey (New York: The Free Press, 1972), 72.

sufficient. The more detailed approach of the physicist would only waste time and money, the critical constraints of modern engineers.

Engineering science is put into play within a realm of applied knowledge. It includes applied science as well as various kinds of conceptualized practices, which are applied as maxims or descriptive regularities. Engineering science is concerned with artifacts, whereas natural science is concerned with nature. Layton shows how engineering science grew not only out of science but also out of heuristic technical practice. By the end of the nineteenth century, the engineering sciences consisted of a complex and integrated system of knowledge, spanning the gamut from highly abstract sciences to collections of procedures collected in various engineering handbooks. Some engineering sciences, like the strength of materials and hydraulics, were based directly on science and were often seen as branches of physics, whereas others, such as the kinematics of mechanisms, according to Layton, evolved from engineering practice. "In either case, their development involved the adoption by engineers of the theoretical and experimental methods of science, along with many of the values and institutions associated with their use. By 1900 the point of origin made little difference; the engineering sciences constituted a unity."[36]

ENGINEERING DESIGN

Engineering sciences serve engineering design processes. Engineering design is concerned with real working devices, structures, organisms, networks, and systems. These are the complex products that result from the processes of engineering. But the notion of design I am emphasizing here refers to the *process* and not so much to the *outcome,* which I discuss later in the book. Engineering sciences such as electrical circuit analysis or solid mechanics deal with idealized versions of these engineered products (although less idealized than physics). Design, engineering science, and basic sciences like physics represent a hierarchy of progressive abstraction connecting the world of engineered artifacts to the ideal world of theoretical physics.[37]

Design, then, is connected with science. But it also involves *art.* In the

36. See Layton, "Mirror-Image Twins," 567–68.
37. Edwin T. Layton, "American Ideologies of Science and Engineering," *Technology and Culture* 17, no. 4 (October 1976): 695.

premodern era, art and technology were not distinct categories; people spoke of the art of farming, the art of medicine, the art of building. In the modern era, art and science or technology have become specialized into their separate sectors of concern and cease to have much to do with each other. Yet a legacy remains. Layton insists that

> design might well be considered a science, and engineers sometimes so treat it, but it is also clearly a matter of art as well. Indeed, it is the oldest part of engineering knowledge to be recorded; the early engineering and machine books are in the nature of portfolios of design, and there is a deep kinship between engineering design and art, running back to the artist-engineers of the Renaissance and earlier. The natural units of study of engineering design resemble the iconographic themes of the art historian. It is no accident that some of the best work on the history of engineering design has been done by historians of art, architecture, and building.[38]

Art in the finished product that results from engineering design is a certain aesthetic quality of appearance in the product. But this element nowadays is for the most part controlled by the advertising industry and moves to the periphery of engineering. Another sense of art, art within the design process, becomes synonymous with the elements of intuition, inspiration, and creativity. An essential aspect of the profession of engineering is the creative activity. Samuel Florman maintains that regardless of the dimension of the engineering design task, whether performed in a team or by an individual working alone, from the ordinary to the ingenious, engineers have some opportunity to create. The creative aspect of the design process is the primary task of the professional design engineer.[39]

But despite kinship with art and the obvious essential creative element in design, in the more modern treatment of engineering, stimulated by advances in systems analysis, there has arisen a rather vigorous attempt to reduce design to a mathematical science.[40] In the premodern era, design was primarily intuitive, grounded in the lifeworld of everyday

38. Ibid., 698.

39. Samuel C. Florman, *The Existential Pleasures of Engineering* (New York: St. Martin Press, 1976), 142.

40. See Layton, "American Ideologies," 698.

involvements. Moving into the modern era, the abstraction out of the concrete lifeworld gave engineering in general and design in particular a more abstract flavor. The attempt to rationalize and systemize the nonrational, unconscious, intuitive elements in the modern engineering design process—which was initiated in the nineteenth century—has been called by Alfred North Whitehead the "invention of invention,"[41] the invention of the method of invention. In his opinion, in order to understand our modern era, with this new method that entered into life, we should neglect all the details and consequences of change and concentrate on the method itself, the real novelty, which has deconstructed the foundations of the old civilization.[42]

Now, Whitehead's dream has never come true and probably never will. Still, at the core of technology, as Mitcham points out, there exists a desire to transform the *heuristics* of technique—problem-solving strategies that reveal solutions without the need to test all possibilities—into *algorithms* of practice—problem-solving methods that exhaust all possible operations and guarantee a solution if one is possible.[43] The tension between heuristics and algorithms is intimately connected with the nature of the ultimate phenomena of matter and energy. Consider the proposal of the mathematician Pierre-Simon Laplace (1749–1827), who insisted that if he were given a complete description of matter and motion at some point in time he could deduce the remainder of the world. Restated technologically, Laplace's postulate might read something like: if we were given a total description of matter and energy at some initial time, we would be able to construct anything we desire. However, Mitcham warns that Heisenberg's uncertainty principle and other aspects of quantum mechanics, as well as contemporary ecological problems, raise fundamental questions about the feasibility of Laplace's postulate and its technological extension.

Actual engineering design typically employs both heuristics and algorithms. It may involve engineering science or it may not. It may involve natural science and mathematics or it may not. It certainly involves a nonreducible element of creativity to varying degrees. Another characteristic that seems pertinent, not only to the design process, but to all engineering

41. Alfred North Whitehead, *Science and the Modern World* (New York: The Free Press, 1967), 96.

42. Ibid.

43. See Mitcham, "Types of Technology," 252.

processes, is a basic intention or will toward efficiency. In ordinary language efficiency means effectiveness. But we must ask, What kind of efficiency are we talking about? And what is this efficiency for? Henryk Skolimowski gives some examples. He argues that "efficiency in surveying is accuracy of measurement; in civil engineering it is durability of structures; in mechanical engineering it comes out as the mathematical ratio of physical energy output over physical energy input, with mechanical engineering always striving for a value of one."[44]

However, Skolimowski's views concerning efficiency are not without controversy. He insists there are patterns of thinking in engineering that are specific to each branch of engineering and independent of contextual considerations. I. C. Jarvie points out that the aim of efficiency for civil engineers is not necessarily durability of structures: "The engineers who devised the Bailey Bridge and the pontoon bridge were given civil engineering problems in which *speed of construction* was far more important than durability."[45] The fact is that deliberations in the social lifeworld can affect the notion of engineering efficiency, even though the de-worlding of the modern engineering project was supposed to disconnect engineering from these contingencies. Mitcham suggests another problem with Skolimowski: he fails to properly distinguish efficiency and efficacy. "The two terms are not interchangeable, as they are often treated, nor is efficiency 'a measure of effectiveness.' A less efficient but more powerful bomb could easily be more effective than a more efficient but less powerful one."[46] Both notions of efficiency and efficacy take their full measure only out of the conversation of the lifeworld. Lacking a full measure, we often make do with solutions that are satisfactory but imperfect. At this stage, our discussion of engineering design points toward the conversations inherent within engineering professionalism and beyond that, not only to the problem of the eventual and irretrievable engineering impact on the human lifeworld, but also to the problem of how conversations among citizens of the lifeworld help shape the engineering design process.

44. Henryk Skolimowski, "The Structure of Thinking in Technology," in *Philosophy and Technology,* ed. Carl Mitcham and Robert Mackey (New York: The Free Press, 1972), 42–49. Electrical engineering and chemical engineering use the same idea of efficiency as mechanical engineering; however, each uses a different notion of energy.

45. I. C. Jarvie, "The Social Character of Technological Problems: Comments on Skolimowski's Paper," in *Philosophy and Technology,* ed. Carl Mitcham and Robert Mackey (New York: The Free Press, 1972), 51.

46. See Mitcham, *Thinking Through Technology,* 226–27.

One approach to design that highlights efficiency and also gives reign to the creative aspect of design comes from the Polish school of praxiology, which I mentioned earlier. Praxiology takes as its task the development of a science of efficient action. It views design as a specific type of efficient action. The work of W. Gasparski has been instrumental in the adaptation of praxiological design theory to engineering design. He has combined heuristic and algorithmic methods into a hybrid design model. Using this method the individual designer can influence creative control over the numerous subtasks involved. The analytical aspects of separating a task into its constituent parts or elements to form subtasks can be reduced to algorithms. This is where computer-aided design (CAD) is beginning to play an ever-expanding role. But the synthetic design aspect of putting parts or elements together into a plan, scheme, or outline requires human creativity. Design, as the old engineering adage goes, is always more than analysis. And what exactly is design? Gasparski provides one of the simplest definitions of the design process: design is the procedure intended conceptually to prepare a change.[47] The MIT Committee on Engineering Design goes into a bit more detail: "Engineering design is the process of applying the various techniques and scientific principles for the purposes of defining a device, a process or a system in sufficient detail to permit its physical realization."[48] Or as T. T. Woodson puts it, engineering design is "an iterative decision-making activity to produce the plans by which resources are converted, preferably optimally, into systems or devices to meet human needs."[49] Most generally speaking, the process of design requires a human being interacting with the world for a particular purpose. If that world is unsatisfactory, design aims to make available to it a product or products that make the world satisfactory. Design makes the unavailable available.

G. L. Glegg identifies three typical departments in a company devoted to engineering design. The *project department* is responsible for new ideas. The *development department* clothes these ideas with mechanisms. The *production department*—if the mechanism proves successful—takes over

47. W. Gasparski, "A Praxiological Theory of Design," in *Praxiological Studies,* ed. W. Gasparski and T. Pszczolowski (Boston: D. Reidel, 1983), 281.

48. MIT Committee on Engineering Design, "Report on Enginering Design," *Journal of Engineering Education* 51 (April 1961): 647.

49. T. T. Woodson, *Introduction to Engineering Design* (New York: McGraw-Hill, 1966), 3.

to refine and streamline.[50] He maintains that these three categories of industrial organization broadly correspond to the three types of engineering design thinking: the inventive, the artistic, and the rational. Their integration is intrinsic to the process of engineering design. The inventive, by itself, however, is often contrasted with design. Mitcham argues that "invention refers to a process which proceeds by nonrational, unconscious, intuitive, or even accidental means. Invention is, as it were, accidental design—and as such highlights the element of insight which plays an important role even in highly systematized design."[51]

Now, engineering design does not just begin spontaneously. It must be activated by needs or desires, which are typically entangled in the life-world and wrapped up with environmental and other concerns. It is generally agreed that engineering design is a multistage process involving iteration and feedback to and from the stages. It starts with a need or desire, draws on available resources, and after several stages like problem formulation, negotiation of constraints, employment of standard forms, generation of possible solutions, optimization, and so on, yields a design that—via a production process, which itself may require an extensive design effort—brings forth a system, device, structure, network, or organism that meets the given need or desire. There are several graphical depictions of iterative design schemata available in the literature.

I offer two cases to illustrate the design process. One involved a minimal amount of engineering science, and the other a great deal of engineering science. Vincenti provides the first example: "In the early 1930's most metal airplanes of American origin were held together by rivets with dome-shaped heads protruding beyond the external surface of the aircraft. A decade later almost all such airplanes had rivets flush with the surface."[52] Why and how did such a change take place? Obviously it constituted an advance in aircraft design. The new design was an important development in the aircraft industry, an industry that was depending more and more on engineering science. Yet, the new design proceeded with hardly any engineering science in the modern sense. Heuristic knowledge and commonsense skills involving trial and error or simple parameter variations were involved in gradually changing the routine design to eliminate

50. G. L. Glegg, *The Design of Design* (London: Cambridge University Press, 1969), 24.
51. See Mitcham, "Types of Technology," 245.
52. Walter G. Vincenti, "Technological Knowledge Without Science: The Innovation of Flush Riveting in American Airplanes," *Technology and Culture* 25 (July 1984): 540.

protruding rivets and replace them with rivets which were flush with the surface. New production techniques were devised and allowable strengths of metals were empirically determined. Yet, as Vincenti points out, "no scientific theorems were called for, and few mathematical equations appeared in the articles and reports, and then only for elementary engineering calculations. A good deal of analytical thinking *was* evident, but such thinking is not solely a province of science. One looks in vain in the story of flush riveting for anything that could seriously be identified as scientific activity."[53]

As another example, consider the design of an autopilot feedback control system that employs a very conceptual and abstract engineering design methodology based on optimization theory. In this type of design, we assume the model of the system, consisting of the dynamics of the aircraft to be controlled, is represented by a set of differential equations. The goals of the design are defined by a value function, which is to be maximized. It is usually an integral or summation function of variables internal and external to the system. The goal might be to achieve maximal comfort for air passengers by minimizing sudden changes in altitude. The dynamics of the model constrains the maximization of the value function, and there may also be additional constraints on other variables. The optimization problem, according to Herbert A. Simon, one of the early pioneers in the application of optimization theory, is to find an admissible set of values of the control variables, compatible with the constraints, which maximize the value function.[54] These kinds of design problems are almost entirely mathematical. A very extensive literature has developed in the past fifty years in mathematical optimization theory covering in detail the best approaches to these kinds of design problems as well as a large variety of related problems.

Both of these kinds of design should be familiar to most engineers, even if these engineers are not "design engineers" per se. Design is more and more being emphasized in engineering education, even though the majority of course work is still focused on the engineering sciences. One issue that often gets shortchanged in most engineering programs is the relation of design to the lifeworld. Design is viewed as a process, as we have stressed in the preceding discussion, and that process can affect and be

53. Ibid., 569–70.

54. Herbert A. Simon, *The Science of the Artificial* (Cambridge, Mass.: MIT Press, 1969), 60.

affected by the human lifeworld. Most engineers, since they have contact with the real world, know that only in the ideal case can the engineering design process be considered to be context-free.

ENGINEERING PROFESSIONALISM

The processes of modern engineering, at least in the last one hundred years or so, are or ought to be carried out in a professional manner. That means that modern engineers should hold themselves to standards that exceed what the law, the market, and ordinary morality might demand.[55] What is this disposition of professionalism that is or ought to be intrinsic to all practices of modern engineering? What does it mean to have a professional disposition toward one's engineering work?

In 1980, George Sinclair lamented the decline in engineering professionalism, claiming that modern graduates from engineering schools were receiving an excellent background in engineering science, but virtually no training in how to perform as professionals in the field of engineering.[56] Sinclair's lament still rings true a quarter of a century later. In spite of tremendous advances in engineering education, not much has changed as far as inculcating a disposition of professionalism is concerned. Since classroom teaching in university settings is typically set up to convey information of the clear and distinct variety, engineering science is normally well covered, engineering design less well covered, and engineering professionalism hardly covered at all. Though engineering professionalism certainly contains distinct features, it seems to be more of an attitude than an easily conveyable body of information. The proper approach for university education might be to teach professionalism not in a separate course, but rather to integrate into already existing courses a concern for the elements of professionalism.

According to Robert Perrucci and Joel Gerstl the four essential dimensions of any profession are *knowledge, autonomy, obligation,* and *commitment.*[57] These are not just features that are present or absent in a

55. Michael Davis, *Thinking Like an Engineer* (New York: Oxford University Press, 1998), 37.

56. George Sinclair, "The Decline of Professionalism in Engineering," IEEE *Transactions on Education* E-23 (November 1980): 177.

57. Robert Perrucci and Joel Gerstl, *Professions Without Community: Engineers in American Society* (New York: Random House, 1969).

given activity aiming to be a profession. They are variables whose value or measure depends on circumstances.

Concerning knowledge, modern notions of engineering professionalism advocate a solid general education program along with a need for broad-based, hands-on technical knowledge as well as the specialized focused knowledge of one's discipline.

Concerning autonomy in engineering, though engineers are protective regarding the integrity of their personal expertise, the modern engineering practice is increasingly a group enterprise. Identification with a group makes the profession of engineering different from most other professions. The individual engineer's autonomy tends to be transferred to a group that is typically further integrated into the business aims of the company. As Layton puts it, "The role of the engineer represents a patchwork of compromises between professional ideals and business demands."[58]

The dedication and commitment of the professional engineer points to the connectivity of engineers to each other and to their profession. The dedication and commitment of the professional engineer provides the point of departure from which issues in engineering ethics can be brought into relief. The first ethical issue of serious concern that the professional engineer may have to face entails the dual obligation to society and to clients. The client may be an individual but is most often the company for whom the engineer works. Does society take precedent over the client or vice versa? This issue has been widely debated in the arena of professional engineering ethics. Is it true that the primary responsibility of a professional engineer is to the client and not to society, and that whistle blowing is incompatible with engineering professionalism? These are important issues and deserve the attention of all engineers. Today in the twenty-first century there is a general sense that an engineer's primary responsibility *is* to society and that whistle-blowing *is* permissible, though only in very complex conflict situations. Charles Harris, Michael Pritchard, and Michael Rabins discuss criteria that might be useful for determining when whistle-blowing is appropriate.[59] The mutual linkages among industry, academia, and government lead to a certain consistency of obligation and commitment of

58. Edwin T. Layton, *The Revolt of the Engineers* (Cleveland: Case Western Reserve University Press, 1971), 5.

59. Charles E. Harris, Michael S. Pritchard, and Michael J. Rabins, *Engineering Ethics: Concepts and Cases*, 2nd edition (Belmont, Calif.: Wadsworth/Thomson Learning, 2000), 197–99.

behavior within the structures of these organizations. Yet society forms the context or backdrop against which these organizations function, and the needs of society may be at odds with the desires of the client.

Codes of ethics, among other things, indicate how the engineering professional is supposed to relate to society. Codes will be looked at more extensively in the next chapter when I investigate process ethics, the kind of ethics appropriate to the modern engineering enterprise. Here is Deborah Johnson's take on codes of engineering ethics:

> Codes may be seen as statements to the public of a commitment to behave in ways that promote social goods and do not harm individuals or society. They may be aimed at providing guidance to members of the profession or as a sensitizing or socializing device. As well, they may be a mechanism for protecting engineers against employers; that is, an engineer might point to a professional code to support his or her refusal to do something an employer has requested. In any case, a review of the codes of professional conduct suggests that engineers have responsibilities to society, employers, clients, and their profession.[60]

The disposition of professionalism, in accord with the dimensions of *knowledge, autonomy, obligation,* and *commitment,* as I have mentioned, can be encouraged by a strong program in basic engineering and a solid broadly based education in the liberal arts. It would also be essential to take to heart one's engineering code of ethics and to acquaint oneself with some of the specific problems of engineering ethics that might arise in one's field of expertise. Investigate for instance case studies, which are widely available in textbooks and on the Internet. Developing some sense of independent thought or autonomy, even though it is invariably transformed by dialogue with other minds, is important to the disposition of professionalism. Equally important is a satisfaction in one's work and a commitment to it as well as a commitment and obligation to clients, colleagues, and to the public in the human lifeworld. That same lifeworld contextualizes the engineering enterprise. And the engineering enterprise, in turn, produces

60. Deborah G. Johnson, "Engineering Ethics," in *The New Engineer's Guide to Career Growth and Professional Awareness,* ed. Irving J. Gabelman (New York: IEEE Press, 1996), 167.

engineered products, devices, systems, and procedures that will eventually impinge upon and no doubt colonize that lifeworld in a variety of ways.

CONCLUSIONS

The process of modern engineering is a series of progressive and interdependent steps that result in a designed product or at least in the design itself. The product might be a tangible artifact like a system, device, network, structure, or organism, or the less tangible plans whereby one of these is to be produced. Engineering design lies at the heart of the process of modern engineering. However, early in the modern era, the engineering sciences began to be developed and increasingly came to inform the process of modern engineering. Modern engineering relies on engineering science, so much so that in contemporary times engineering practice that does not rely on the engineering sciences to some extent is viewed as craftwork or premodern engineering, not modern engineering.

The third primary element that characterizes the process of modern engineering, along with engineering design and engineering science, is engineering professionalism. The professional spirit that guides the process of modern engineering pervades the entire modern engineering enterprise, including engineer, engineering, and the engineered. As far as the person is concerned, the focus of Part II, a professional character is indicated by a committed and caring attitude that includes honesty and objectivity. As far as the process is concerned, the focus of Part I, a professional design is indicated by a disposition that strives for a high level of technical excellence and is attuned to one's professional code of ethics. As far as the product is concerned, the focus of Part III, a professional attainment is indicated by the creation of products that meet the needs of the customer and that harmonize with potential end-users and their lifeworlds.

TWO

PROCESS ETHICS

In this chapter, I discuss the ethics of the modern engineering process. Concern with professionalism mandates an ethical component within the enterprise of engineering. Most professions, engineering included, address that mandate via the institution of codes of ethics. In the previous chapter, I considered engineering professionalism from the perspective of Perrucci and Gerstl's four dimensions of *knowledge, autonomy, obligation,* and *commitment.* It is specifically the dimension of obligation that creates the tie between the process of engineering and the social lifeworld. Professions are *obliged* to serve the common good.

As I discussed in the previous chapter, on the one hand, within the field of modern engineering, decontextualized procedures tend to predominate, but on the other, the thrust of professionalization within the enterprise connects engineering processes to the human lifeworld. When engineers ignore context, their engineering processes can run along smoothly, simply, in an almost algorithmic fashion. And sometimes that does no harm. But as professionals we know reality is more complex. In fact, immersed within it we encounter the ambiguity—specifically with regard to the extent to which we need to factor contextualizations into our engineering practice—which heightens the interest and challenge inherent in the work of the engineer. Broadly put, disconnection or de-worlding, at least in theory, has constituted an integral feature of engineering since the dawn of the modern era. Connections or contextualizations, though still a fact of life within modern engineering, played a larger role in the premodern engineering endeavor and will again play a significant role within the focal engineering venture. But within the modern engineering enterprise, practice often proceeds by trying to sidestep context. A concern for environmentalism, for example, is typically reduced to satisfying minimal EPA guidelines, which do not require an engagement with deep ecology or any kind of holistic environmental thinking. Still, we know that connections, like the impact of engineered products on the human lifeworld, do exist, and within the professional nature of the engineering enterprise is where they must be dealt with.

The duties, rights, and obligations of the engineer are summarized in codes of ethics. The cynic would contend that if one is truly honest and responsible, then one has no need for a code of ethics, and if one really needs a code of ethics, it would not do one any good anyway.[1] Then there is also the opinion that codes of ethics include a hodgepodge of considerations, like matters of etiquette, moral ideals, and duties. Often also they have been used if not intended to protect professional privilege.[2] However, I will affirm the positive contribution of codes of ethics to the development of engineering professionalism, stressing the entrenched professional paradigm, which specifies the primacy of responsibility to society. Each branch of engineering has a code, and these codes are all very similar. These codes are primarily based on process ethics of the deontological and utilitarian varieties, though there are also elements of virtue ethics involved, as well as other types of ethical systems.

The realm of theoretical and universal ethical judgments is distinguished from the realm of the more everyday concrete particular ethical judgments. The latter are usually based on the character of the individuals involved and will be stressed in Part II. The philosophy of Immanuel Kant can provide a foundation for the realm of universal ethical judgments, within which the idea of duty stemming from pure reason is paramount. Consequentialism is another ethical theory that is at home in the realm of universal ethical judgments. Utilitarianism is the most familiar form of consequentialism. I will look at both the Kantian and the consequentialist conceptual schemes and their applicability to engineering ethics.

In addition, I will consider the *aims* of process ethics. I suggest the aims of health and safety, environmental sustainability, and social justice and will look at specific examples of engineering practice reaching or failing to reach those goals. These examples will ground the discussion and give content to otherwise abstract ethical methodologies. Deliberations within the

1. John Ladd, "Collective and Individual Moral Responsibility in Engineering: Some Questions," *IEEE Technology and Society Magazine* 1, no. 2 (June 1982): 3.

2. Vivian Weil, "Action and Responsibility in the Engineering Profession," *Center for the Study of Ethics in the Professions Occasional Papers*, no. 2 (August 1979): 9.

realm of theoretical and conceptual ethical discourse are empty without a grounding in the concrete realm of real-world engineering practice, just as deliberations within the realm of praxis need the guidance provided by concepts generated from within the conceptual realm.

PROFESSIONALISM AGAIN

A problem of the engineer in attaining recognition as a professional by society has historically centered around the issue of social responsibility. Edwin T. Layton, in tracing the rise of American communities of scientists and engineers, points out some problems the field of engineering had in achieving professional stature. The scientific community was striving for *disinterested* knowledge, while the engineering community aimed at altering the world in certain ways and not others, in the service of the *interests* of its employers and clients. Scientists who worked primarily in academia were generally less beholden to government and industry. As Layton showed, the engineers who enjoyed independence like the scientists usually lacked necessary prestige to lead their profession. "This had the practical effect of giving the control of the engineering profession to men who were linked by ties of self-interest to those who were using, and in some cases, misusing technology."[3] Nevertheless, over time, the idea of a disinterested science has faded, as the interests of science come to match the technological interests of industry and the government. Today, instead of science as such or technology as such, we often speak of technoscience.

In the early days of the profession, the rank-and-file engineer was not considered a professional in the service of society because of this need to serve the corporate whim. This attitude, though mitigated, still exists, although circumstances have changed considerably. Today, for one thing, the corporation is more socially responsible. This is largely due to an expanded awareness on the part of society, which now insists that corporations play by the rules. That includes an appeal that government enforce laws currently on the books. Consumer advocacy, for instance, has demanded the engineering profession be accountable for its actions as well. Also, today academia is more dependent on industry and government than in the past. Most academics and their institutions today could not survive without the grants they receive from government and industry.

3. Layton, "Mirror-Image Twins," 579.

Engineering schools, in turn, recognize their obligation to graduate students who are properly prepared for work in the world of industry. Within the complex web of these interactions, what often gets overlooked is the leadership and vision that engineering schools are supposed to provide for the industries they serve. To be socially responsible, engineers should be taught, like medical doctors, to "first do no harm." Then by working through their professional societies, and by acquiring a broad-based general education employing critical academic skills, students can learn to be leaders and help shape a vision of a better world. Amplifying the dimension of social responsibility would no doubt go a long way toward augmenting the professional status of all engineers.

Other views of professionalism, which focus more directly on engineering professionalism, are provided by Robert G. Sargent and Charles A. Pratt.[4] Sargent believes that the two most important ingredients for engineering professionalism are ethics and competence.[5] Pratt believes education in a specialized field of endeavor and in the liberal arts and sciences as well as guidance by a recognized and well-publicized code of ethics and some form of certification are sufficient.[6] These views are in essential agreement when we recognize that certification assumes a certain level of competency, which is generally acquired in academia. Many branches of engineering have forms of certification that require the candidate pass an examination or have a degree from an accredited school of engineering. The main issues seem to reduce to ethics and education, with the former being the focus of this chapter.

Ethics in engineering presents a tangled complex of problems. Chief among these is the issue of social responsibility. Though codes of ethics are proliferating among the various branches of engineering, ethical problems seem to be proliferating as well. The Institute of Electrical and Electronic Engineers (IEEE) provides a code that members are encouraged to follow. According to the IEEE code, and like most other engineering codes, the engineer, among other things, should be protective of the safety, health, and welfare of the public. But codes of ethics, it seems, can only provide general guidelines; solutions to actual ethical problems can only come from conversations among actual human beings.

4. Robert G. Sargent, Charles A. Pratt, et al., "Ingredients of Professionalism: A Panel Discussion," *Proceedings of the 1985 Summer Computer Simulation Conference,* July 1985.

5. Ibid., 170.

6. Ibid., 171.

What often makes ethical problems seem intractable is the inadequacy that many engineers feel regarding ethical issues even when armed with a code of ethics. As pawns in the game of corporate whim, many engineers feel that they are powerless. Though "powerless" may be a bit of an overstatement, at times engineers may feel caught in a maelstrom,[7] to use philosopher John Ladd's expression, and think they can do little about a situation that they deplore. As W. M. Evan puts it:

> One reason for the widespread neglect on the part of engineers of the problem of social responsibility for technological change is the difficulty of accepting responsibility for events over which they exercise virtually no control. As salaried employees performing, in the main, a staff function, engineers are rarely in a position of making policy decisions concerning the wisdom of developing or not developing a new engineering product, or concerning what, if any action, might be taken to counteract its potential or actual negative social effects.[8]

The end-use of a product is the purpose to which it is to be put. In his discussion of end use in engineering, C. Thomas Rogers maintains that, for the most part, when engineers discuss it, their critique is rather rudimentary, because generally the engineer is involved with enabling clients to achieve their objectives within technical and other constraints. In point of fact, whether they like it or not, engineers are often means to others' ends.[9]

To be more than a functionary, the professional engineer must assume some responsibility for the end-use problem, particularly those problems where ethical issues are at stake. Weapons work comes immediately to mind. What about the ethics of the atomic or hydrogen or cluster bombs? What about land mines? Land mines keep on killing long after the reason for emplacing them has dissolved into history. Is it ever really ethical to design a land mine?

Appropriate engagement with ethical issues in engineering in general or with end-use problems in particular presumes a morally responsible

7. See Ladd, "Collective and Individual Moral Responsibility," 4.

8. W. M. Evan, "The Engineering Profession: A Cross-Cultural Analysis," in *The Engineers and the Social System*, ed. R. Perrucci and J. E. Gerstl (New York: John Wiley & Sons, 1969), 124–25.

9. C. Thomas Rogers, "The Ethical End-Use Problem in Engineering," *Research in Philosophy and Technology* 8 (1985): 77.

engineer. To be morally responsible, the professional engineer must cultivate concern, care, and foresight. To extend this notion of moral responsibility, Ladd discusses collective moral responsibility: "In as much as one person's being responsible does not relieve others of responsibility, everyone in a group may have moral responsibility for a certain thing."[10] Responsibility for safety, for example, belongs to all those involved in the process of making available any particular commodity. Since engineers mainly work on teams, if the team as a whole shoulders the burden of responsibility, then the previously discussed sense of individual powerlessness can be substantially mitigated.

Rogers indicates that engineers could be more responsible and have a better framework for dealing with the end-use problem if the codes of ethics were made more reasonable and practical. Rogers states his position in opposition to Samuel Florman, who believes that engineers are essentially moral. "Florman allows that there have been excesses in technology—even excesses abetted by engineers—but he holds that these anomalies should be controlled by regulation and law (formal social means) rather than by 'limiting' individual engineers as with codes of ethics."[11] In fact, these social controls that Florman refers to often arise too late, after the damage has been done. Social laws and regulations are necessary but not sufficient. How can the public be more adequately served? What engineers seem to mean when they speak of the needs of society is really an array of individual and corporate interests that may not include any representation of the public interest.[12]

Now, even though engineers, like most other professionals, believe that having a code of ethics is a major part of being a professional, it turns out that codes of ethics tend to be ignored in making ethical decisions. This was strikingly illustrated by results from a survey conducted in 1980 by the editors of the trade journal *Chemical Engineering,* who presented a set of nine hypothetical cases studies illustrating some ethical dilemmas that arose in engineering practice. Included in these cases were situations in which an engineer became aware that his employer was illegally dumping chemical wastes, and another in which an engineer was asked to "fudge" performance test data for a new product. The cases were accompanied

10. See Ladd, "Collective and Individual Moral Responsibility," 9.
11. See Rogers, "Ethical End-Use Problem," 89.
12. Ibid., 90.

by a questionnaire in which readers were asked to indicate what course of action an "ethical" engineer should take in each case. Of the large number of responses (4300) to the survey, fewer than half a dozen even mentioned the Code of Ethics of the American Institute of Chemical Engineers. The readers tried to resolve each problem in a very individual and personal manner. The result was a diversity of opinions concerning the best solution to each problem. Although this kind of ethical relativism might satisfy Florman, who believes in the inherently high moral character of engineers, many feel that there is a need for the engineering profession to make its codes of ethics relevant to real situations encountered by real engineers. The professional codes should support the ethical conduct of its members, so that individual engineers need not feel they must grapple with these ethical dilemmas in a solitary and isolated way.[13]

Let us assume that Florman is right to insist that engineers are inherently moral. It would still do no harm to support this inclination with a strong and relevant code of ethics. The Accreditation Board for Engineering and Technology proposed as a fundamental canon for a code of ethics that "engineers shall hold paramount the safety, health, and welfare of the public in the performance of their professional duties."[14] Such statements, while setting the right priorities, leave open a lot of questions, such as Who is the public? What is the public welfare? Who determines it? And how? As mentioned earlier, such statements seem to be capable only of providing general guidelines. Rogers advocates more strongly worded codes, in particular, the changes proposed by Stephen Unger, Andrew Oldenquist, and Edward Slowter. The previously mentioned fundamental canon could be strengthened as follows: "Engineers shall regard their responsibility to society as paramount and shall . . . endeavor to direct their professional skills toward conscientiously chosen ends they deem, on balance, to be of positive value to humanity; declining to use those skills for purposes they consider, on balance, to conflict with their moral values."[15] Such a provision, Rogers maintains, would help to provide professional engineers with the resources needed to more adequately deal with matters of public interest.

13. Rosemary Chalk, "Ethical Dilemmas in Modern Engineering," *Technology and Society* 9 (March 1981): 1.

14. See Rogers, "Ethical End-Use Problem," 86.

15. Ibid., 97.

KANT AND DEONTOLOGY

Heinz Luegenbiehl examined three general categories of engineering ethics, namely, professional ethics, business ethics, and technological ethics.[16] He concluded that some combination of these might constitute a proper groundwork for an engineering ethics and would indicate obligations to society, to employers or clients, and to the engineering profession. Which obligation takes precedent, however, is not always obvious, though the tendency today is to give most weight to society. According to Kant, one ought to do one's duty, regardless of the particular obligation that arises. But what is that? To act in accord with respect for the law. But which law? Kant tells us in the maxim he calls the Categorical Imperative: "I ought never to act except in such a way *that I can also will that my maxim should become a universal law.*"[17] This version of the famous Categorical Imperative of Immanuel Kant, according to Simon Blackburn, is called

> the Categorical Imperative in its first form, the so-called Formula of Universal Law. Later on Kant glosses it in other ways. One is "Act as if the maxim of your action were to become by your will a *universal law of nature*" (the Formula of the Law of Nature). Another, possibly the most influential, is "So act that you use humanity, whether in your own person or in the person of any other, always at the same time as an end, never merely as a means" (the Formula of Humanity). It is not at all clear that these different versions can be derived one from the other, but Kant regarded them as somehow equivalent.[18]

What might the Categorical Imperative mean for engineering ethics? Let's say I adopt the principle that I should never take bribes. Can I will as a universal law that no one should ever take bribes? Yes, I think that would serve the profession of engineering well. This does indeed seem like a good idea, although in some cultures gift giving is common practice, which raises the question of when does a gift become a bribe? In

16. Heinz C. Luegenbiehl, "What Is Engineering Ethics? A Prolegomena," *Science, Technology & Society Newsletter of the Lehigh University STS Program,* no. 36 (June 1983): 1.

17. Immanuel Kant, *Groundwork of the Metaphysic of Morals,* trans. H. J. Paton (New York: Harper & Row, 1964), 70.

18. Simon Blackburn, *Being Good* (Oxford: Oxford University Press, 2001), 120–21.

general, many factors may be involved in evaluating a gift, such as its size, the timing, and the intent of the giver. All these factors may need to be weighed in order to arrive at a specific determination in a specific case, but that does not deny that bribery is wrong and should not be tolerated. The integrity of the engineering enterprise depends on adherence to imperatives like this. As Blackburn puts it: "The most persuasive examples of the Categorical Imperative doing some real work are cases where there is an institution whose existence depends on sufficient performance by a sufficient number of people."[19] Engineering would certainly be one of those institutions.

Now the portion of Kant's philosophy that is summarized in the Categorical Imperative is often referred to as his deontology. What does that mean? "The term 'deontology' derives from the Greek words *deon* (duty) and *logos* (science). Etymologically, it means the science of duty. In current usage, however, its meaning is more specific: A deontological theory of ethics is one which holds that at least some acts are morally obligatory regardless of their consequences for human weal or woe. The popular motto 'Let justice be done though the heavens fall' conveys the spirit that underlies most deontological theories."[20]

Another possible maxim based on Kant's deontology might be that all engineers should tell the truth. If I tell the truth because it is convenient, because I fear getting caught, or because I believe I will be rewarded for doing so, then my act is not generally morally worthy. But if I do so because I recognize that I must respect the other person, then I act from duty and my action is right.[21]

When we think of Kant, we usually think of abstruse theoretical philosophical ideas. But Kant also wrote much about character. And character is what the virtues exhibit. It's just that Kant's notion of character is rather conceptual. He believed that the first effort we should put into moral education was to establish a character, which G. Felicitas Munzel, in explicating and translating Kant, defined as the *accomplished ability to act in accordance with maxims*.[22] But maxims, articulating moral law,

19. Ibid., 121.

20. *Encyclopedia of Philosophy*, ed. Paul Edwards (New York: Macmillan, 1967), s.v. "deontology."

21. Deborah G. Johnson, *Computer Ethics*, 2nd edition (Upper Saddle River, N.J.: Prentice Hall, 1994), 29.

22. G. Felicitas Munzel, *Kant's Conception of Moral Character* (Chicago: University of Chicago Press, 1999), 61.

have no value unless they are followed. The theoretical must be made to accord with the actual. Objective practical reason must be made subjectively practical. Moral law is enacted in the lifeworld by a person of good moral character, one who practices the virtues. As Kantian thinking would have it, the objective and subjective dimensions of "moral law and character may be understood as two sides of one whole, with the second dependent on the first for its supreme principle, but the first also dependent on the second for its actuality."[23]

UTILITARIANISM

First formulated by Jeremy Bentham (1748–1832) and J. S. Mill (1806–73), utilitarianism considers the end of moral behavior to be the attainment of the greatest good for the greatest number. A utilitarian could countenance sacrificing one person to save fifty. A Kantian who had adopted the principle that the individual was sacrosanct, however, could not. Thus, while Kantian morality is deontological, utilitarianism is *consequentialist*. It can most broadly be described as the view that the rightness or wrongness of an action is determined by the goodness and badness of its consequences. The consequence of greatest interest to utilitarianism is generalized benevolence or happiness. "The basic principle is this: *Everyone ought to act so as to bring about the greatest amount of happiness for the greatest number of people.*"[24]

Many varieties of utilitarianism exist, including act and rule utilitarianism, egoistic and universalistic utilitarianism, hedonistic and ideal utilitarianism, and normative and descriptive utilitarianism. I will look only at the first pair and consider whether engineering ethics should focus on individual acts or rules of behavior. Rule utilitarians believe one should posit rules, or embrace already extant rules, the following of which would maximize happiness in the general sense and in the long run. "The rule utilitarian does not consider the consequences of each particular action but considers the consequences of adopting some general rule, such as 'Keep promises.' He adopts the rule if the consequences of its general

23. Ibid., 70.
24. *Encyclopedia of Philosophy*, ed. Paul Edwards (New York: Macmillan, 1967), s.v. "utilitarianism."

adoption are better than those of the adoption of some alternative rule."[25] Act utilitarians believe one should act in such a way as to maximize hapiness in the general sense and in the long run. They understand actions to mean "particular actions" and must assess the rightness or wrongness of each individual action directly by its consequences.[26]

Consider the case of Mr. and Ms. Jones, who were both hired directly out of college as engineers by a famous Silicon Valley electronics company. Mr. Jones discovers something interesting about the new semiconductor material that the company has started to use for their circuit boards. It emits trace amounts of the toxic gas phosgene, especially in the assembly stage of production. It is still within EPA limits and would not be illegal, but if word got out, the company's reputation would suffer. He feels he should eventually switch to a different material, which, unfortunately, cost twice as much. He has already shipped three orders and has kept quiet about the toxicity problem. Mr. Jones decides to discuss the problem with his wife before telling anyone. He thinks the company should quietly make the changes and not rock the boat. Recalling the three shipped orders would entail a big loss. If he keeps quiet, no one will ever know. Mr. Jones has already taken his ethical stance. He is an act utilitarian. He figures that the greater good is better served by not talking than by broadcasting the problem and maybe forcing a recall. No one using the product will be hurt that much but with a recall the company could lose several thousands of dollars.

Both Mr. and Ms. Jones are members in good standing of the IEEE. They have its code of ethics on the walls of their respective offices. The first tenet states very clearly that we do agree *to accept responsibility in making engineering decisions consistent with the safety, health and welfare of the public, and to disclose promptly factors that might endanger the public or the environment.* Unlike her husband, Ms. Jones is a rule utilitarian, and she takes that rule of the IEEE very seriously. She feels that in following that rule she would engender the greatest good for the greatest number of people. Her ethical stance is to call for disclosure. The dangers, though slight, are real. The health and safety of the public must take precedent over the company losing money.

In general, act utilitarians treat rules as guidelines that can be ignored

25. Ibid.
26. See Harris, Pritchard, and Rabins, *Engineering Ethics,* 147.

when it becomes obvious that more happiness will result from breaking them. On the other hand, rule utilitarians take rules to be binding. If rules are justified, then acts that violate or ignore those rules are wrong.

THE AIMS OF THE ENGINEERING ENTERPRISE

The usual view of the aims of the engineering enterprise is fairly clear. An engineer should strive to meet a client's specifications, to receive fair compensation, and to extend the state of the art in his or her field of expertise. He or she should design a product to be useful, durable, simple, and reliable. But how do these aims serve not just the engineer's personal good or the corporate good, but the common good? If the moral good is to be served, then the engineer must take as paramount the health, safety, and welfare of the general public.

Health and safety are obvious aims of the processes of the modern engineering enterprise. The notion of human welfare is a bit more complex. We can agree, however, on two points. Without social justice for all there can be no welfare for the general public. Without protecting the environment, all other aims and goals that the engineering enterprise might strive for are diminished. Thus I take as crucial the goals of *social justice,* starting at the local level and spreading out to incorporate global concerns, and *environmental sustainability,* starting globally and pointing into our local concerns.

Process ethics should oblige engineering, at the very least, to guarantee that its products are not unsafe and the health of consumers and end users is not compromised. Like doctors, engineers should first do no harm. Process ethics should further oblige engineers to value ecological sustainability and to strive for social justice with their products. Other moral goals may surely be posited and pursued, but health and safety, social justice, and environmental sustainability are fundamental.

Health and Safety

We can associate the value of health and safety with the *personal* dimension. Each and every individual, under the umbrella of respect for persons, deserves the best possible consideration regarding matters of health and safety in the commonweal of their engineered worlds.

Engineers have a distinct obligation to protect the health and safety of the public. There are risks involved with every engineering venture, but the thoughtful engineer can reduce them by increasing safety factors and by making, whenever possible, a risk analysis to determine whether or not the risks will be acceptable. William W. Lowrance defines "risk" as "a compound measure of the probability and magnitude of adverse effect."[27] Risk can thus be objectively measured as a product of a likelihood and relative magnitude of the harm that could be done. Assume, for instance, that the magnitude ranges from zero to one hundred, so a value of ten would be a small magnitude and one of ninety a large magnitude. Thus, according to Harris, Pritchard, and Rabins, "A relatively slight harm that is likely might constitute a greater risk than a relatively large harm that is far less likely."[28]

The idea of risk can be subsumed by the broader notion of "cost." Cost-benefit or risk-cost-benefit analyses are increasingly common practices in corporate engineering. A certain confusion arises, however, because engineers often lump together the moral with the non-moral costs. The risks associated with a product, like a new television, are part of the overall cost of buying and using such a product. I not only pay the $399 for the product, but I also tacitly agree to endure a risk of radiation, I pay the psychic cost of "dumbing down" of family and friends, and I run the risk of becoming a television addict or a couch potato. The health and safety associated with any engineered product, then, can be incorporated into the notion of cost, broadly construed, where the notion of health needs to encompass not just physical but also mental, emotional, and spiritual health. Recognizing that getting a purchase on some of these ideas can be a strenuous task, engineers, with the products they engineer, should at least strive, as best they can, to maximize the health and safety of end-users subsequent to their taking up with those products. Or, thinking in terms of costs, engineers should strive, as best they can, to minimize the unhealthy and dangerous aspects of their engineered works.

Nowadays, public agencies generally scrutinize large-scale engineering projects, and routine repetitive projects typically follow well-known standards of health and safety. For example, the U.S. Department of Labor administers the Occupational Safety & Health Administration (OSHA).

27. William W. Lowrance, "The Nature of Risk," in *Societal Risk Assessment: How Safe Is Safe Enough?* ed. Richard C. Schwing and Walter A. Abers Jr. (New York: Plenum Press, 1980), 6.

28. See Harris, Pritchard, and Rabins, *Engineering Ethics*, 155.

The mission of OSHA is to assure the safety and health of America's workers, among other things, by setting and enforcing standards in the workplace. But often not covered by routine investigation are novel or low-volume products. When they are being manufactured or later being used, there may be side effects that are not widely known, that may be difficult to recognize, and whose origins are hard to trace. Examples include cases of products that either incorporate toxic chemicals or use them in their manufacture. Engineers aware of the hazardous nature of such products and processes, even without an OSHA-type intervention, have the responsibility either to remove hazards of which they are aware, or to alert those who could be subjected to any remaining hazards, and then to initiate procedures for the safe disposal of toxic waste.[29] Of course, accomplishing these goals may require painstaking effort, especially if one acts alone. It may be effective to form into groups of all those who are involved in the engineering of a given novel or low-volume product. Again, as John Ladd has suggested, for some things, like safety, everyone may be responsible, and collective responsibility lightens the load for the individual engineer.

Social Justice

The value of social justice is naturally and initially associated with the *social* dimension. The human lifeworld in its community aspect is the domain of this concern. Once the health and safety of the end-user of a product is adequately taken into account, then we should ask about how society in general might fare, how it might fare well, as a result of engaging with this product. This brings to the discussion the idea of social justice, the second of three values the modern engineering enterprise ought to champion. Social justice is a broad and open-ended notion. First of all, what do we mean by society? According to philosopher John Rawls, "society is a more or less self-sufficient association of persons who in their relations to one another recognize certain rules of conduct as binding and who for the most part act in accordance with them."[30] Social

29. Roland Schinzinger, "Technological Hazards and the Engineer," in *Social, Ethical, and Policy Implications of Engineering,* ed. Joseph R. Herkert (Piscataway, N.J.: IEEE Press, 2000), 136.

30. John Rawls, *A Theory of Justice* (Cambridge, Mass.: The Belknap Press of Harvard University Press, 1971), 4.

justice or social equality refers to the right of all persons within society to have access to resources they need.

Of course, what is fair for one may be unfair for another. A necessary resource for me may not matter for you. What are needs for some are mere desires for others. All these contentions should be put on the table for us to deliberate about in the conversation of the lifeworld. Regarding social justice, Rawls says: "For us the primary subject of justice is the basic structure of society, or more exactly, the way in which the major social institutions distribute fundamental rights and duties and determine the division of advantages from social cooperation."[31] The basic inequalities inherent in any and all societies must be dealt with by the principles of social justice. However, given the diversity of cultures alive in the world today, arriving at a universally acceptable concept of social justice is problematic. Questioning the concept is not only proper but imperative, insists David Harvey, because "too many colonial peoples have suffered at the hands of western imperialism's particular justice, too many African-Americans have suffered at the hands of the white man's justice, too many women from the justice imposed by a patriarchal order and too many workers from the justice imposed by the capitalists, to make the concept anything other than problematic."[32]

Nevertheless, in keeping with the aims of process ethics, engineers are responsible for bringing products into the world that contribute to some sense of social justice. These products at least should not advance injustices among human who take up with them. Here, again, is the idea that first one should do no harm. In discussions of social justice, the respect for persons mandated by Kant's ethical principles should be paramount. We ought not to sacrifice any single person for the greater good. This means that we must reject utilitarian ethical judgments, at least act utilitarian judgments. The controversial nature of the utilitarian position was strongly brought forth when it was invoked by T. J. Rogers, CEO of Cypress Corporation, in his exchange with Cypress shareholder and Franciscan nun, Sr. Doris Gormley. Rogers maintained that societal good was the greatest good for the greatest number, which was the greatest economic good, which was equivalent to corporate profit maximization. But is it really true that the greatest good for the greatest number is the

31. Ibid., 7.

32. David Harvey, *Justice, Nature & the Geography of Difference* (Malden, Mass.: Blackwell Publishers, 1996), 342.

economic good, and the greatest economic good is corporate profit maximization? These assertions are contestable. Gormley questioned them and proposed that Cypress needed a more diverse board of directors with a broader view of societal concerns who could keep the larger question of the common good before the mind of the corporation and its management.[33] The real issue is whether maximum profit translates into social justice. Should corporations serve only their shareholders, or should they also strive for a broader sense of social justice?

With regard to the social justice issue, perhaps corporate capitalism can embrace a Kantian rather than utilitarian form of ethics, especially as far as respect for persons is concerned. But this would necessitate an "impure" brand of capitalism, because "pure" capitalism encourages *homo economicus* to seek only his or her own interests. Pure capitalism, as Eva Bertram and Kenneth Sharpe point out, encourages investors to relocate businesses to optimize returns, even if that means a decline of neighborhoods and communities. It encourages corporate officers to streamline operations and lay off workers. It invites consumers to comparison shop for the best deals, even if that means abandoning the corner drugstore for Costco and Wal-Mart.[34]

The problem with an economy in the grip of the capitalist "take" on reality is that everything becomes commodified and human relationships become purely functional and instrumental. An attitude of respect for persons becomes more and more difficult to maintain. It seems that pure corporate capitalism is beyond the reach of not only utilitarian ethics but also Kantian ethics. Further evidence of this can be found in looking at the conditions of "perma-temps" (long-term temporary hires). The firm that hires perma-temps has no legal responsibility for their social security or health insurance, or long-term investment in their training and professional development. Absence of employer loyalty and commitment make employee loyalty and commitment irrational. As Bertram and Sharpe

33. The discussion between Gormley and Rogers took place in the *Wall Street Journal* in July 1996 and was analyzed in a piece published on the Web by the Center for Ethics, Capital Markets and Political Economy, "a non-profit organization established in 1994 to provide a discussion forum and information resource for persons who believe that moral concerns should be taken into account in economic and political thinking" associated with the University of Virginia. See http://www.iath.virginia.edu/cecmpe/gormley.html/.

34. Eva Bertram and Kenneth Sharpe, "Capitalism, Work, and Character," *American Prospect*, September 11, 2000, 1. This article is available on line at http://www.prospect.org/print-friendly/print/V11/20/bertram-e.html/.

explain, "the new economy teaches workers to be 'career entrepreneurs,' assuming nothing from their employers, always preparing to move on, seeing coworkers instrumentally as stepping stones to new jobs. *U.S. News and World Report* recently headlined an article 'Why It Pays to Quit.' The article's subtitle explains, 'Loyalty, shmoyalty. In today's frenzied job market, staying put gets you nowhere. Walking out gets you ahead.'"[35]

If we are reduced to treating others as stepping-stones and if utilitarianism is just a mask for corporate greed, then where does ethics fit in the new economy? And, in particular, what can engineers do about their responsibility for attending to social justice? Well, fortunately, corporate laissez-faire free market capitalism is just an abstraction, perhaps not even an ideal, especially in light of its tendencies toward some not very healthy outcomes, like expanding the gap between the rich and the poor. The machinations of the processes of capitalism are grounded in the human lifeworld. That ground can be overlooked, as is the tendency within the modern engineering enterprise, but it cannot be willed away.

A society generally consists of a body of people occupying some specific territory and organized under a specific government. Typically, such a society is called a state. "Since the seventeenth century, political philosophers have been largely preoccupied with the relations of the state and the individual, with the citizen's rights, if any, against the state, with the right of the state to punish, to promote morality, or to regulate the affairs of other associations such as families, trade unions, and churches."[36] The state, at least in modern democratic societies, while demanding certain responsibilities of its citizens, is supposed to guarantee certain freedoms for its citizens. Such guarantees can soften the blow of capitalism's insatiable will to power. According to Sam Gindin, capitalism's moral limits and political vulnerability were apparent from the beginning, raising issues of an ameliorative response.[37] Gindin shows how the French aristocrat Montesquieu, writing in the mid-eighteenth century, expressed a remarkably early argument for the welfare state as a counter to the realities of capitalist society. The state owes every citizen, in Montesquieu's own

35. Ibid., 5.

36. *Encyclopedia of Philosophy,* ed. Paul Edwards, s.v. "state."

37. Sam Gindin, "The Terrain of Social Justice" (http://www.socialjustice.org/pubs/gindin.pdf), 5. This essay is a revised version of the inaugural lecture for the Packer Endowment in Social Justice, originally delivered April 17, 2001.

words "an assured subsistence, proper nourishment, suitable clothing, and a mode of life not incompatible with health . . . whether it is to prevent the people from suffering, or whether it be to prevent them from revolting."[38]

However, realizing the modern welfare state has always been a challenging process, especially since the ascendancy of John Locke's idea of the primacy of the individual. With the idea of individualism firmly entrenched in the mind of the modern American and European, taking care of the needy tended to be secondary. The *will to empower* of social justice movements struggled with the *will to power* of capitalist enterprise. That will to power, Gindin pointed out, thwarted questioning by social democracy of the social relations at the heart of the economy, the political division of society into those who led and those who followed, and the divide embedded in a welfare state between those who planned and organized social services and those who were dependent on them.[39]

But the will to empower has had its day as well. The welfare state flourished throughout the mid-twentieth century from the social programs of FDR in the 1930s up until the 1960s and 1970s. In recent years, with tightening economies around the world, capital (those in control of capital) has begun to pull in the reins on the welfare state. The concessions it had granted to the needy came to be understood as having undermined the market precisely because these concessions, for instance, aid to dependent mothers, diverted too many resources to unprofitable ends.[40] In effect, capitalism is colonizing the state, or colonizing society by declaring war on the welfare state.

For now the question is, What role can engineering play in serving the spirit of social justice, wrapped as it is in the inequalities capital demands? Freedom or freedoms are at stake. Social freedoms coupled with individual freedoms link strongly with social justice. A particular kind of individual freedom, namely market freedom, links strongly with capitalism. When the free market idea becomes dominant, social justice is seen to be antagonistic to freedom. The underclass will remain the underclass as long as its only freedom is the freedom to purchase this engineered commodity rather than that one. As Gindin puts it:

38. This is a reference in Gindin's essay quoted from G. C. Morris, "Montesquieu and the Varieties of Political Experience," in *Political Ideas*, ed. David Thompson (London: Penguin Press, 1990).

39. See Gindin, "Terrain of Social Justice," 6.

40. Ibid., 6.

> For a minority, economic freedom revolves around the power to organize production and accumulate; for the rest, freedom to sell one's productive potential in a labour market and, on the basis of that, to exercise some personal choice in consumer markets. What the minority is accumulating as part of its freedom includes power *over* the labour of others and therefore over their "individuality." The freedom/power to sell one's productive potential and to exercise some choice in consumer markets, in contrast, is founded on a dependency on those who provide the jobs and the commodities available for consumption.[41]

Social justice aims at establishing a little more equality, both in the processes of social organization and control, and in the results that accrue from those processes. Unions, for instance, try to improve the conditions of workers and the benefits that flow from those improvements. Engineers, however, are generally conditioned to pay little heed to the problems of social justice. Make a good product, put it into the world, try to make a buck, help to keep the company solvent. But this hyperpragmatic attitude belies the professionalism that is supposed to permeate the engineering enterprise. There was and is the notion of advancing the welfare of the public. If we take seriously the ethical mandate to promote social justice with the products we engineer, then we still need a more concrete sense of social justice. Again, how to get a purchase on a viable notion of social justice?

Perhaps abstract theories of social justice are not what we are after. Perhaps we want a more direct source of how to be toward social justice. David Harvey discusses the work of Iris Marion Young, whose endeavors have been in the field of the politics of difference. Young has proposed a family of conditions relevant to a contemporary conception of social justice. She has identified "five faces of oppression":

1. *Exploitation* (the transfer of the fruits of the labor from one group to another, as, for example, in the cases of workers giving up surplus value to capitalists or women in the domestic sphere transferring the fruits of their labor to men);

41. Ibid., 4–5.

2. *Marginalization* (the expulsion of people from useful participation in social life so that they are "potentially subjected to severe material deprivation and even extermination");
3. *Powerlessness* (the lack of that "authority, status, and sense of self" which would permit a person to be listened to with respect);
4. *Cultural imperialism* (stereotyping in behaviors as well as in various forms of cultural expression that "the oppressed group's own experience and interpretation of social life finds little expression that touches the dominant culture, while that same culture imposes on the oppressed group its experience and interpretation of social life"); and
5. *Violence* (the fear and actuality of random, unprovoked attacks, which have "no motive except to damage, humiliate, or destroy the person").[42]

Freedom from these five faces of oppression would be a good start for a way of being with social justice.

Another take on the social justice issue is to look at human rights. The United Nations Development Program publishes a Human Development Report in which they listed several universal freedoms of human rights. Among these are freedom from discrimination, from want, from fear, from exploitation, to develop one's own potential, and to participate in decision making.[43]

Certainly, the full extent of what each of these entails requires a lengthy conversation. Nevertheless, these freedoms, if achieved, would be indicative of a just society. To aim their designs at achieving social justice, engineers might work, like Stanford mechanical engineering graduate Martin Fisher, to empower Kenyan farmers by developing low-tech leg-powered irrigation pumps. Fisher's efforts have created thousands of jobs for impoverished Africans. From an article by Jane Ellen Stevens, in the *San Francisco Chronicle*: "He has a dream that tilts at windmills: to help create an African middle class. Develop some low technology in a world with hardly any and you'll tap into the entrepreneurial spirit that exists everywhere, says Fisher. Then watch as millions of small businesses grow

42. See Harvey, *Justice*, 349.

43. From the *Human Development Report 2000* of the United Nations Development Program (UNDP), chap. 2; available at http://www.undp.org/hdr2000/english/HDR2000.html/.

and thrive. Today in Africa, where 10 percent of the population is rich and the rest is poor, the idea of a middle class is a wildly divergent concept."[44]

Developing the potential of disenfranchised peoples should certainly shrink the gap between the rich and the poor, and these are processes in which engineers can participate. But Martin Fisher is not an average engineer. Most engineers work within a corporate structure and cannot go running off to a third world country to develop a bright idea. They are rooted in particular local worlds, they have family responsibilities, and they owe some sense of loyalty to their company.

Engineers do not need to uproot and travel to distant shores to contribute to a more socially just world. They need only do what they can as engineers to change their products or influence company policies in a way that augments social justice. A company, for instance, might design and manufacture cell phones. But the materials it uses are available cheaply only from a small country in South America. By buying from them, it turns out, the company is supporting a repressive dictatorship. That company would augment social justice if it were convinced by its engineers to purchase its materials elsewhere. Engineers are also citizens of their communities. Most of the issues that are discussed in the conversation of the lifeworld are, to varying degrees, and especially these days, technologically based. The contributions of engineers to these issues can be invaluable. Engineers as specialists can advance the cause of social justice by a judicious communication of technical information to the lay public. Explaining and interpreting technical issues can be an important service in the achievement of social justice.

Freeman Dyson suggests that solar power, genetics, and the Internet can be brought into the service of social justice. Solar energy, he maintains, is most available where it is most essential, namely, in tropical latitudes, where most of the world's population lives, rather than in temperate latitudes.[45] The problem, certainly, is that solar energy is still too expensive. Dyson believes a new technology is needed that would combine photoelectric systems with energy crops. He says we can develop crop plants which would not need to be harvested and which could convert sunlight

44. Jane Ellen Stevens, "Martin Makes a Middle Class," *San Francisco Chronicle*, December 8, 2002, (http://techawards.thetech.org/news_press/2002.12.08.sfgate.com.pdf), 2/11.

45. Freeman Dyson, "Technology and Social Justice," in *Society, Ethics, and Technology*, ed. Morton E. Winston and Ralph D. Edelbach (Toronto: Wadsworth, 2000), 145.

directly into fuel. An energy crop could be a permanent forest. Trees could convert sunlight to liquid fuel and deliver the fuel directly through their roots to a system of underground pipelines.[46] Of course, the trick here is to have the people in these tropical areas—which is where most of the world's poor live—benefit from these technologies. How to keep global capital from moving in and developing the energy sources and colonizing the locals? That is a question that the conversation of the lifeworld must engage. Genetic engineering is an essential tool with which to realize this dream of a solar energy landscape. Dyson predicts that in a few decades "we shall have achieved a deep understanding of the genome, an understanding that will allow us to breed trees that will turn sunlight into fuel and still preserve the diversity that makes natural forests beautiful."[47] These are some tasks for energy engineers and genetic engineers that might advance the cause of social justice.

Dyson also advocates pushing the Internet into remote locations to allow businesses and farmers to participate in the burgeoning global economy. The Internet would allow people in remote places to make connections. They could make business deals, buy and sell, keep in touch with their friends. They could continue their education. They could pursue their hobbies and avocations, with full knowledge of what was going on in the rest of the world.[48] By working toward closing the digital divide, engineers would be advancing the cause of social justice. Whether or not Dyson's vision is attainable, he does point in a direction toward which the modern engineering enterprise could be reoriented if it is to seek to fulfill its ethical obligation to serve social justice. Again, we cannot be naive about the profit-motive mandate of the engineering enterprise. The goal of engineering in the embrace of the capitalist framework is not to produce systems, devices, networks, structures, and organisms—that is, commodities—as such, but to produce profits and surplus labor. How that goal can be merged with the goal of social justice has no easy answer, but the conversation must continue.

A final example of an engineered reality contributing to social justice is the story of the two Lees, Lee Felsenstein and Lee Thorn. They were instrumental in bringing the benefits of the digital age to the mountainous jungles of northern Laos. They assembled a group of computer hot-shots to

46. Ibid., 146.
47. Ibid., 147.
48. Ibid.

install a bicycle-powered computer Felsenstein created from off-the-shelf odds and ends. "They call the invention the Jhai Computer, Jhai meaning 'hearts and minds working together.' It was built because the villagers asked Thorn for a way, any way, they could better tap into their country's economy and have contact with the outside world."[49] The immediate consequence of this project will likely be enhanced connectivity and communication between the two hundred villagers of Phon Kham, who are some of the neediest people on the planet, and the rest of the towns in Laos. Social justice will be augmented by the people in Phon Kham being able to find out what the market is like in the big towns where they sell their merchandise. Then they can charge a fair market price, instead of the below-market price they had been forced to sell at previously.

Environmental Sustainability

The value of environmental sustainability is initially associated with *global* concern. The reach of environmental concern recognizes that the modern engineering enterprise impacts the entire planet. Now, if the health and safety of a given end user of a given product is adequately taken into account, and if that product is seen to contribute to social justice or at least not to promote injustice, then we should ask about how the environment in general might fare, how it might fare well, as a result of engaging with this product.

An immediate problem one encounters when bringing a notion like environmental sustainability under the umbrella of theories of process ethics is that these theories as originally formulated were about how humans should treat one another. Nonhuman entities like nature were not considered. However, our primal intuitions tell us that we should not harm others and that includes animals and nature generally. We are all connected and belong to the earth, from which we arise and to which we return. The voice of nature speaks to us and reveals, in many cases, especially today, an ill or traumatized environment. As part of the natural system of life, our primal intuitions direct us to aim for harmonious accord with human life and nature in general. Ethical theories, then, can perhaps be broadened to acknowledge all living things.

49. Kevin Fagan, "Pedal-powered E-mail in the Jungle," *San Francisco Chronicle*, January 17, 2003, 3/6; available at http://www.sfgate.com/cgi-bin/article.cgi?file=%2Fchronicle%2Farchive%2F2003%2F01%2F17%2FMN86676.DTL/.

For instance, cutting down the redwoods in the Pacific Northwest provides a livelihood for local loggers but destroys animal habitats. Loggers and environmental activists need to find common ground. Some sort of compromise must be hammered out, drawing perhaps on the insights of utilitarianism, Kantian ethics, or some other ethical question. If, for example, environmental sustainability is posited as a value all can agree on, then a conversation might evolve.

In addition to primal intuitions, we must also pay homage to primal experiences. In the intuitions of everyday moral experience, we find that both the consequences of our actions and respect for the subjective integrity of the other, human or not, are morally relevant.[50] The environment therefore is not mute. It speaks to us of its moods. We may experience it as good, as vacant, as beautiful, as powerful, as valuable, as vulnerable, as meaningful. Whether we may take nature as being as important as humans, or as something of which we are an integral part, or as uniquely determinative of our very being, the point is that we treat nature with respect. It is sufficiently significant that, under normal circumstances, we value it as a good and we will it to be sustained. And, of course, without a healthy environment everything ventured within that environment will be flawed. Therefore, environmental sustainability is a reasonable value to strive for.

In advanced capitalist societies, the typical approach to environmental problems is to intervene only after the fact. David Harvey calls this "the standard view" of environmental management.[51] On the other hand, the approach of "ecological modernization" adopts a more dialectical view of social and ecological change. It depends upon and promotes a belief that economic activity inevitably tends to produce environmental harm in the sense of some kind of disruption of nature. Society, then, ought to be more proactive toward these potential disruptions by expanding environmental regulation and ecological controls. An ounce of prevention is preferable to a pound of cure. Harvey elaborates: "This means that the *ad hoc,* fragmented and bureaucratic approach to state regulation should be replaced by a far more systematic set of politics, institutional arrangements, and regulatory practices. The future, it is argued, cannot

50. Charles S. Brown, "The Real and the Good: Phenomenology and the Possibility of an Axiological Rationality," in *Eco-Phenomenology: Back to the Earth Itself,* ed. Charles S. Brown and Ted Toadvine (Albany: State University of New York Press, 2003), 10.

51. See Harvey, *Justice,* 373.

be expected to look after itself and some sorts of calculations are necessary to configure what would be a good strategy for sustainable economic growth and economic development in the long run. The key word in this formulation is 'sustainability.'"[52]

Of course, what we mean by, or what is meant by, sustainability is an open question. The Brundtland Report from 1987, which stems from the UN-sponsored World Commission on Environment and Development (WCED), says sustainable development means that actions taken in the present should not compromise the ability of future generations to meet their needs. The social responsibility of the practicing engineer extends into the future, though how far into the future, as we have indicated, is an open question. Is it necessary, for example, that we leave no footprints? Or is it sufficient simply to follow EPA guidelines?

What sustainability implies is that the value of the total stock of assets, or some measure of the quality of life, must remain constant across the generations. If we destroy or use up all the trees, for instance, but we provide genetically engineered organisms that take in carbon dioxide and yield the oxygen we need to survive, would that be an acceptable scenario? Is it possible to compensate for reduction of the quality of life in one area by with an elevation of it in another? I somehow doubt it.

Trees, for one thing, cannot or should not be reduced to their oxygen-producing capability. A tree is also a source of shade, a refuge from the rain, a wonder to behold, a thing that gathers a world about it. All this gets submerged in the view of the tree as just a resource. We are charged from a point of view of technological utility to look at the tree instrumentally, but as fully engaged humans being-in-the-world to look at the tree in terms its ethical and spiritual values as well.

That we ought to promote sustainability seems incontestable. But exactly what form the idea of sustainability should take is undecided. Certainly we would like to move in the direction of a less unsustainable way of being. First of all, it is necessary to distinguish between maintainable and sustainable. The former means to just keep, for example, a process from going under, and the latter means to more fully participate in that process, to nurture and cherish it, to give life to it, or to prepare the way to let the life it manifests come to fruition. In speaking of the meaning of the idea of sustainability, Aidan Davison argues:

52. Ibid., 377.

> that it is best understood as the ongoing ability to support, relieve, sustain, or nourish. Unlike the adjective *sustainable,* which is applied as easily to the productivity of a munitions factory as it is to the harnessing of solar energy, the verb *sustaining* holds open the actively normative questions that the idea of sustainability raises. We are required to probe: What truly sustains us? Why? And how do we know? Conversely, we must ask: What are we to sustain above all else? Why? And how may we do so?[53]

Davison goes on to distinguish between *cultural* sustainability and *technocratic* sustainability. The former is concerned with strategies for sustaining the sources of sustenance. These are understood predominantly in political, moral, and spiritual terms. The latter is concerned with strategies that are understood predominantly in descriptive or instrumental terms that allow for little evaluative judgment.[54] In light of this distinction how might modern engineers proceed with ethical evaluations based on environmental sustainability? To attain a full view of matters at hand, they must incorporate both notions of cultural and technocratic sustainability. The latter is necessary, but only both together are sufficient. As Wendell Berry maintains, we need to use both the *rational* and the *sympathetic* mind in our dealings with nature.[55] The rational mind is in charge of managing nature in an efficient manner, but the sympathetic mind respects nature in itself and sees it as worthy of our gratitude. A life of consumption is to some degree mandated, but for true sustainability a life of engagement is essential. And to direct our efforts at a sustainable environment requires a very conscious engagement.

Consider, for instance, the recycling movement. It has been all the rage for the past few decades, but has actually been around for hundreds of years. Paper recycling in the United States can be traced back to 1690 in the Rittenhouse Mill near Philadelphia, where used cotton rags and waste paper were remanufactured into usable paper. New York City initiated a curbside recycling program in the 1890s and required residents to separate their refuse into bins for paper, organic materials, and general trash.

53. Aidan Davison, *Technology and the Contested Meanings of Sustainability* (Albany: State University of New York Press, 2001), 63–64.

54. Ibid., 64.

55. Wendell Berry, "Two Minds," *The Progressive,* November 2002.

I remember shortly after the end of the Second World War, when I was very young, the junkman used to come up and down the alleys of Chicago, riding his horse-drawn cart and yelling "RAGS AND IRON" which they gave us money for and which they would then recycle. Modern recycling, however, began in earnest in the early 1970s, when several recycling centers, mostly nonprofits, sprung up around the country following the enthusiasm generated by the first Earth Day in 1970.[56]

The mechanism behind the recycling process is generally straightforward and can be seen as pointing toward *technocratic* sustainability, while *cultural* sustainability comes into play with the political decisions that need to be made to initiate the recycling process in the first place. Convincing people that recycling is a good thing, something we ought to do, requires a conversation and points also toward cultural sustainability. Get people convinced. Initiate public policy discussion. Inquire about how to sustain the sources of sustenance. Make it, for example, illegal to dispose of recyclable plastic in the garbage bin. Our intuition, stimulated by the conversation of the lifeworld, tells us that at the current rate of production and consumption, if we do not recycle, we will soon be buried in our garbage. The involvement of engineers in this conversation concerning sustainability is crucial because they can demonstrate the various methodologies of recycling as well as the advantages and disadvantages of each.

Will the use of genetically modified (GM) foods yield a sustainable form of agriculture? Though of all the different kinds of engineers, the civil engineer is most likely to be involved in a discussion about sustainability through recycling, the genetic engineer is likely to be part of the discussion concerning GM foods and their contribution to the sustainability of the planet's food supply. What about GM foods? Are they "frankenfoods" as their opponents insist? Are they damaging the environment or saving the world? Their advocates are not all just trying to make a buck. Many are true believers. With the global population exploding as it is, we can only feed the masses with some drastic technological fix like GM foods. Anthony J. Trewavas tells us that researchers recently developed a strain of GM rice that has a 35 percent higher yield than unmodified varieties. Trewavas continues:

56. Information about the history of recycling comes from Tufts University; see http://www.tufts.edu/tuftsrecycles/more/USstats.html/.

> Clever plant breeding in the early 60's produced rice and wheat plants with well over double their previous yield; such progress enabled a parallel doubling of mankind, without massive starvation. But this option is now exhausted. Ignoring the problem, leaving billions to starve in misery, the worst of all tortures according to Amnesty International, is not an option either. "Every man's death diminishes me because I am part of mankind; ask not for whom the bell tolls. . ." is a philosophy I know many here will share with John Donne. So where one grain grew before we now again have to ensure that two will grow in the future. Currently GM is our best option to achieve this difficult task.[57]

Unfortunately, the use of GM crops encourages monoculture, which reduces biodiversity. Cross-fertilization between GM crops and unmodified varieties can lead to what opponents of GM foods call "genetic pollution." As Vandana Shiva insists, biodiversity already holds the answers to many problems for which genetic engineers are seeking solutions. If we turn away from the engineering paradigm to an environmentalist one, we will not only conserve biodiversity but also meet our needs for food and nutrition and avoid the risks that GM foods might pose.[58] Another way to say this is that the engineering paradigm needs to be expanded to incorporate the more holistic view of environmentalists. Their view, roughly and briefly, is that the earth is not a resource in the service of the human will to power but something of which we are a part.

Radical environmentalists, as Michael E. Zimmerman explains, consist of deep ecologists, social ecologists, and ecofeminists.[59] These different schools of thought all advocate caution in the implementation of genetic engineering. Such views are usually condemned as being Luddite, but they are rather seeking the proper place for new technologies, like genetic engineering, and seeking a balanced and harmonious human/world constellation. Environmental engineers, genetic engineers, and environmentalists of every kind, along with involved citizens need to discuss the larger issues

57. Anthony J. Trewavas, "GM Food Is the Best Option We Have," in *The Ethics of Food*, ed. George E. Pence (Lanham, Md.: Rowman & Littlefield, 2002), 150.

58. Vandana Shiva, "Genetic Engineering and Food Security," in *The Ethics of Food*, ed. George E. Pence (Lanham, Md.: Rowman & Littlefield, 2002), 146.

59. Michael Zimmerman, *Contesting Earth's Future: Radical Ecology and Postmodernity* (Berkeley and Los Angeles: University of California Press, 1994).

surrounding GM foods and decide whether and how their introduction will satisfy the requirement of environmental sustainability.

BRINGING IT TOGETHER

Consider the engineering of *R*adio *F*requency *ID*entification (RFID) devices. Assume the technical aspect of the engineering meets all the standards of efficiency and productivity. How might we assess the moral dimension of the engineering of RFID devices? We should certainly gauge it against the standards of health and safety, environmental sustainability, and social justice. But where is this assessment to be carried out? And by whom? I suggest we make our first assessments at the professional level. As I noted earlier in the chapter, Heinz Luegenbiehl sees the engineering profession as mediating between individual engineers and society. Initially, we might involve individual engineers along with members of a professional ethics committee on an assessment team. Ultimately, RFID devices will have to be assessed within the conversation of the lifeworld, with all interested parties involved.

To initiate discussions with regard to assessment we use a quasi-mathematical approach. This will not provide an ethical solution but might help to orient the discussion. The procedure is simple but can help to provide a point of departure for deeper discussions. If we assign a value function (J_{hs}, J_{es}, J_{sj}) to each of the three moral standards we are aiming at (health and safety, social justice, and environmental sustainability), we can write an expression for the overall ethical value (J_p):

$$\text{By definition } J_p = \alpha_1 J_{hs} + \alpha_2 J_{es} + \alpha_3 J_{sj}$$
$$\text{with } \alpha_1 + \alpha_2 + \alpha_3 = 1.0$$

The α_i terms are weighting factors whose values are to be determined by consensus. Initially assume all three value functions are equally weighted, so all α_i terms will be set to $^1/_3$. Assume all value functions can range from −3 to +3. A negative value for one of the J functions may indicate that social justice is not achieved. A positive value for another J function may indicate that health and safety are well provided for.

The professional level committee might decide that the health and safety associated with the RFID device is quite well accounted for. In the manufacturing process the usual precautions are taken. The chips are produced using typical processing procedures. Usual safety measures are taken. No problems are foreseen in manufacturing. However, there is the problem of the electromagnetic energy to be beamed at the RFID device from the reader device. Although the power is very low, if there is a huge proliferation of such activity, it is not clear if this presents a danger to people close by. The same uncertainty with cell phones, however, does not seem to have deterred anyone from owning a cell phone. The committee rates health and safety acceptable, but not excellent. The participants assign J values independently of one another and the moderator averages the results, yielding $J_{hs} = 1.5$.

No problems are seen with disturbing the environment with RFID devices. Again, the moderator averages the results, this time yielding $J_{es} = 3.0$. But social justice issues present a problem. There is a worry that RFID devices can and will be used to spy on people. A person's buying profile can be generated, for example, and her privacy can be invaded. Another issue is that a loss of jobs is likely to occur as the efficiency and productivity in the distribution chain rises because of an increase in the number of RFID devices. Generally, the powerless masses might suffer because RFID devices will augment the power of those who already control the flow of information by deciding what choices will be available to the masses. But from the point of view of several of the evaluators, the RFID device appears rather neutral. The moderator averages the results, yielding $J_{sj} = -1.0$. The total value function J_p is computed:

$$J_p = 1/3(1.5) + 1/3(3.0) + 1/3(-1.0) = 1.17$$

which is on the positive side and indicates a pretty good ethical assessment. There are more positives than negatives, but that there are negatives at all indicates that caution needs to be taken. This number of 1.17, again, is not an "answer" to the ethical question about whether or not the modern engineering process involved in RFID device design, development, and manufacturing is good. But it can provide a point of departure for further discussion. It can help to orient the discussion as it moves beyond the professional level into the conversation of the lifeworld.

CONCLUSIONS

The three ethical aims, then, of the engineering enterprise—health and safety, social justice, and environmental sustainability—are noble and necessary values, and achieving them is a worthy goal for the engineering enterprise. To pursue these values is to pursue the good. Whether they are pursued under a framework of Kantian ethical theory, or utilitarianism, or some other theories, these three aims can provide guidance for the process of modern engineering practice.

The actual assessment of the modern engineering enterprise must take place within the profession, among professionals, and through conversations in the lifeworld. The connection of the profession to society has a crucial import. Society grants privileges to the profession in exchange for service the profession renders to society. Assessment conversations must involve the voice of society. How that voice gets a hearing is an open question.

What David Harvey, a social geographer, has to say about the integration of environmental questions with justice questions bears on all aspects of process ethics. He warns us that the integration of environmental questions and social justice questions are too easily absorbed by the dominant forms of economic power, whereas "the environmental justice movement . . . puts the survival of people in general, and of the poor and marginalized in particular, at the center of its concerns."[60] Issues of class, race, national identity, and gender come into the spotlight. Environmental justice brings out the political dimension of the environmental sustainability concern. Treating environmental sustainability or social justice separately ignores the fact that they are bound up with each other. If environmental sustainability is at the expense of the poor or the underclass, then it is not truly aimed at the moral good. And if social justice policies do not take into account the health of the planet, then they are not truly aimed at the moral good either.

60. See Harvey, *Justice*, 386.

THREE

COLONIZATION

Engineering affects the world as the world affects engineering. The contemporary German social philosopher Jürgen Habermas stated that the realm of system colonizes the lifeworld by imposing upon it the values of efficiency and productivity. On the other hand, conversations and decisions made in the human lifeworld contextualize systems, including the engineering project by motivating it and giving it direction. Colonization and contextualization travel the same road but in opposite directions. I contend, however, that the colonization of the lifeworld by engineering takes priority in the contemporary era of the modern engineering enterprise.

Colonization or the imposition of the values of productivity and efficiency on the human lifeworld is certainly not a bad thing, not in any absolute sense, unless it is done inappropriately. Often when a product colonizes the lifeworld, the efficiency it engenders is seen as a wonderful thing. That was undoubtedly the case with a product like the shopping cart. However, the introduction of the automobile, which was welcomed as an efficient new way to get around, has had many unforeseen negative consequences. Perhaps the advantages of enhanced mobility blinded people to them. Perhaps there was a "technological imperative" at work, a certain temptation to push toward the greatest feat of technical performance or complexity currently available. The key issue with colonization is, Who determines, or should determine, when the line of appropriateness is crossed? Who decides how much efficiency we need or whether or not a product will really augment the quality of our lives? This is where the conversation of all involved parties becomes crucial.

I confess that the word *imposition* has a funny ring to it. On the one hand, an "imposition" is a burden. But on the other, it can also refer to the laying-on of hands in a confirmation ceremony or a ritual blessing.

I draw on both of these meanings to imply both a burdening and a disburdenment. Classical colonization, in fact, burdened the colonized and disburdened the colonizers. Colonization is always good for the colonizers. But for the colonized? Who are the colonized? All of us who have had or felt they have had the values of efficiency or productivity imposed on us with or without our consent. Making the colonization of the lifeworld by the engineering enterprise a good—or at least acceptable—thing for the colonized is a central concern of engineering ethics.

TO COLONIZE IS TO AFFECT

To colonize is to *affect* but not necessarily to *effect.* The steam engine certainly affected the world, but did the invention of the steam engine actually *cause* the Industrial Revolution? It certainly influenced it. But the Industrial Revolution had many causes. Lewis Mumford believed the clock was the key to the Industrial Revolution,[1] others say it was the rise of market capitalism, and some say it was culture itself changing that led to the new social configuration we call the Industrial Revolution. The steam engine did not *effect* the Industrial Revolution, although it certainly *affected* it. However, if colonization of the lifeworld by the engineering enterprise proceeds unabated, affecting will become effecting.

In his book *Technopoly* Neil Postmann suggests that history can be segmented into three phases: tool-using, technocracy, and technopoly.[2] The tool-using phase corresponds to the era of the premodern engineer. The technocracy phase corresponds to the modern engineering enterprise, wherein engineering *affects* society with its increasingly bountiful catalog of devices and measures, the promise of technology realized. In the technopoly phase the engineering project *effects* a complete colonization of the lifeworld by systems.

The more complete technopoly becomes, the more the engineering project expands to fill the lifeworld (Figure 2). Fortunately, we are still in the technocracy phase of history, but if we are to keep technocracy from becoming technopoly, if we are to keep the totalizations of productivity and efficiency at bay, we will have to counter colonization with contextualization. And that is the goal of the focal engineering project—

1. Lewis Mumford, *Technics and Civilization* (New York: Harcourt Brace, 1934).
2. Neil Postman, *Technopoly: The Surrender of Culture to Technology* (New York: Alfred A. Knopf, 1992).

achievement of a proper balance between the forces of colonization and contextualization through the creation of products that will not just do no harm in the world, but will actually contribute to a consensually arrived at notion of the good.

Borgmann shows how devices proliferate in our modern technological society. They relieve us of one burden after another, but at the same time they affect and colonize the lifeworld, sometimes in negative ways. I do not currently own a cell phone, but my daughter brought hers along on our last family trip, and having the capability to call ahead and make arrangements and reservations while speeding down the interstate proved to be extremely useful. Now I am thinking about getting one for my car.

Devices emerging from the engineering enterprise affect the nonsystemic lifeworld. They colonize it by lending efficiency and productivity to tasks that may have been considered onerous. All around us we see engineered devices impacting, affecting, disburdening, colonizing, and congesting our worlds. We certainly live in a technocracy, in Postman's sense of the term, but it is not yet a technopoly. To keep technopoly at bay and to relieve the congestion in our lives, what might we do? Since it is clear that overcolonization is what is at stake, it seems that we need a revival of the contextualizing response. Focal engineering, seeking to balance colonization forces with forces of contextualization, participates in public policy discussions within the conversation of the lifeworld.

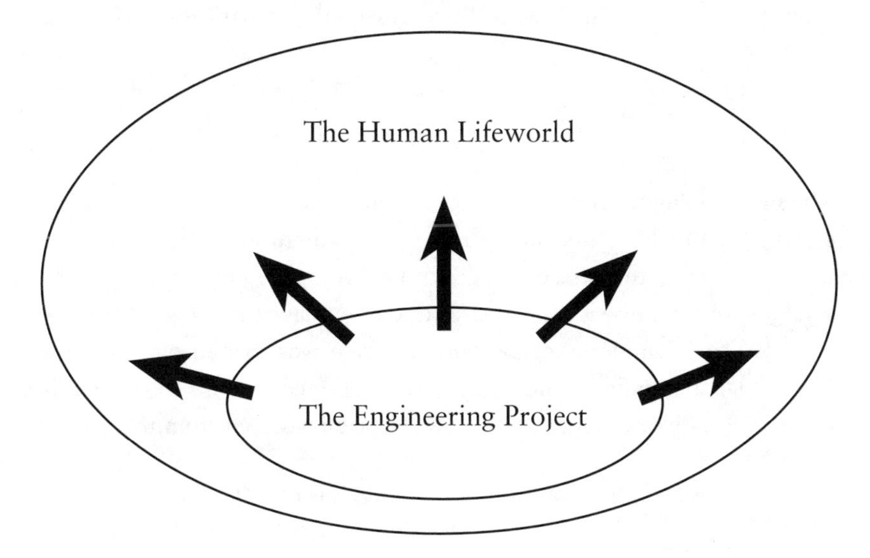

Figure 2. The Technopoly Phase

A machine is type of device that certainly affects the lifeworld. Thomas Misa has studied how machines make history, how these particular engineered products transform the lifeworld. He shows, through an extensive review of the literature, "that those historians (and others) adopting a 'macro' perspective are the ones who allow technology a causal role in historical change. They deploy the Machine to make history. This causal role for the machine is not present and is not possible in studies adopting a 'micro' perspective."[3] Misa points to several macro or big-picture cases that are relevant to my concerns in this chapter on colonization. For instance, several historians write about the rise of electricity and gas utility services in the early twentieth century.[4]

Once the populace had been convinced of the necessity of having these services, wherever the utility companies put up the power lines and laid the gas lines, there it was that the people went and communities "naturally" developed where these services were close by. The technology of the electrical power lines, product of the engineering enterprise, had a distinct effect on the shape and placement of urban communities. Utilities colonize the lifeworld. But it is important to note here that the actual devices—the generators, the transformers, and the wire—were all parts of a larger technological system. Devices emerging from the engineering enterprise affect the nonsystemic lifeworld, but they are usually mediated by realms of technological systems and systems in general. The typical flows of impact or influence can be represented schematically (Figure 3).

TRIPLE COLONIZATION

The modern engineering enterprise colonizes the realm of technological systems, the realm of technological systems colonizes the realm of systems in general, and systems in general colonize the lifeworld. At every level, of course, there is some degree of contextualization, but its voice is muted. The premodern engineering endeavor was carried out within a contextualizing milieu of intimacy between project and lifeworld. Modern engineers work alone in tiny cubicles, isolated even from their team

3. Thomas J. Misa, "How Machines Make History, and How Historians (and Others) Help Them to Do So," *Science, Technology, & Human Values* 13, nos. 3&4 (Summer & Autumn 1988): 308.

4. Ibid., 316.

members except when meeting to discuss issues pertinent to their projects. The result is a sharp division between project and lifeworld.

Decisions made within the engineering enterprise yield products that impact the realm of technological systems and colonize it by incorporating into these technological systems with increasing strictness the values of efficiency and productivity while reducing the relevance of context. A particularly pertinent example is the laptop computer. Now as powerful as any desktop system, the laptop has extended the reach of the computer networks already present in almost all technological systems. When work stations are mobile, the context in which work occurs no longer matters.

Technological systems, in turn, impact the realm of systems in general. There is an increasing demand nowadays for practical application and commercial exploitation in the global market place of any and all research results in basic science. One consequence of these pressures is that any new findings within the realm of system, especially scientific findings, are quickly integrated into the realm of technological systems. Another is that scientific research itself is inclined away from increasing knowledge for its own sake and toward increasing knowledge for the sake of its potential usefulness. Technoscience colonizes science. The military, for example, is always anxious to adapt new developments in technoscience

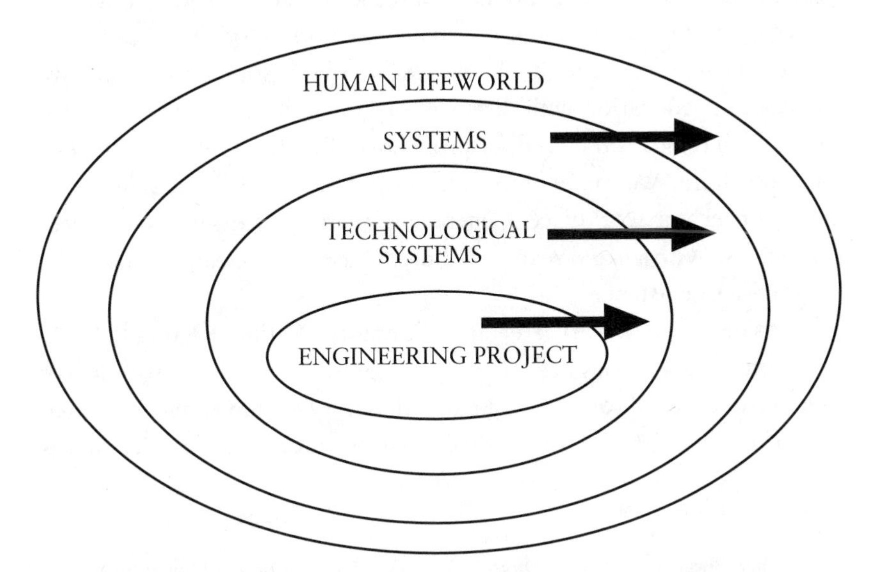

Figure 3. How the Colonization Effect Is Mediated

to warmaking. Technoscience colonizes the military. Medicine eagerly adopts new drugs, tools, and procedures to control and maintain the natural systems of the human body. Technoscience colonizes the body. And our small-scale, natural system of growing indigenous crops in tune with regional peculiarities and seasonal cycles is being replaced by technology-dependent agribusiness supplying world markets with highly processed and genetically engineered foodstuffs. Technoscience is colonizing agriculture.

Finally, systems in general colonize the nonsystemic lifeworld. Whenever a situation arises that requires a functional and pragmatic approach, engagement itself is colonized. I and Thou become Its. And wonder is usurped by systemic amazement.

Thus, the colonization of lifeworld by the engineering project, especially in modern times, is ongoing. Contextualization of the engineering project by decisions made in the human lifeworld, though muted within the modern engineering enterprise, does surface from time to time in processes like "enrolling" products—to use Bruno Latour's term[5]—in the natural environments in which they function. Context can then come to light. But these efforts to constrain modern engineering practice are generally feeble or easily sidestepped wherever there is a firm commitment to the reengineering of the environment to suit the needs and desires of a consumer lifestyle based in a global market economy. The pervasive spirit of engineering takes more and more of the ground once held by nonengineered systems, or even the nonsystemic lifeworld. Today, we can engineer political processes with scientific polls and statistical predictions. We can engineer education with distance learning and computerized education, which removes the need for professors. We can engineer our investment portfolio. We can engineer our attitudes. With robotic devices we can engineer our workplace. Genetic engineers are working to engineer our babies. With virtual reality and digital avatars we can engineer our very sense of existence.

Consider the IMAX theater at Yellowstone National Park. It has a screen that is six stories high. This virtual presentation of the wonders of Yellowstone is said to surpass the real thing.[6] Once the systemic simulation of reality replaces reality, then what becomes of the reality that is

5. Bruno Latour, *Science in Action: How to Follow Scientists and Engineers Through Society* (Cambridge, Mass.: Harvard University Press, 1987).

6. Albert Borgmann, *Holding On to Reality: The Nature of Information at the Turn of the Millennium* (Chicago: University of Chicago Press, 1999), 217–18.

supposed to ground, condition, and orient the realm of system? Will the uncolonized lifeworld shrink to nothing as systems dominate more and more of our everyday human activities?

The use of generalized engineering theories and procedures further separates modern engineers from the concrete lifeworld and distinguishes them from premodern engineers, who relied on grounded intuitions and experience. In comparing the theories of knowledge relevant to ancient and modern technology, Carl Mitcham says that "the former relies for guidance primarily on sensorimotor skills, technical maxims, and descriptive laws, whereas the latter uses these resources plus technological rules and theories."[7]

The premodern engineering endeavor was embedded in the lifeworld without sharply delineated boundaries between technological system, system in general, and lifeworld. But the contextualization of the engineering project by the decisions of the lifeworld was stronger then than is the case today. There is a kind of inverse proportionality at work here. The dimmer the boundaries of contextual embeddedness, the more forceful the process of contextualization tended to be. Today with the sharp boundaries, we find the modern engineering enterprise to be largely decontextualized. We can passively allow that to continue, or we can start talking things over. As TyAnna Lambert maintains:

> Habermas' establishment of system as an enemy of the lifeworld leads to paradox, for if through communicative action, the individual can effect change and make choices for development of the social structure, the systems that drive society will also be changed. A lifeworld, which empowers the individual through communicative action, if taken to its pragmatic conclusion, would help the individual to contribute to the creation of a system that provides a basis of societal structure. The system, then, would derive from the lifeworld of communicative action, rather than inhibiting or "colonizing" it.[8]

Colonization can be countered by contextualization. This triple colonization seems to be part and parcel of the orientation of the modern

7. Mitcham, *Thinking Through Technology*, 207.

8. TyAnna Herrington Lambert, "Jürgen Habermas: Luddite Dragon or Defender of the Weak?" 4–5; available at http://www.daedalus.com/didak/cccc95/tyanna.html/.

engineering enterprise. The premodern engineering project, which I am calling the engineering endeavor, was oriented more toward contextualizing than colonizing. The obvious thing to do, in pursuit of a reform of the modern engineering enterprise, would be to strengthen the forces of contextualization.

HYPERMODERNISM

In his book *Crossing the Postmodern Divide* Borgmann maintains that if modernism continues to develop unchecked it will develop in the postmodern era into its logical extension, which he calls *hypermodernism*. The hypermodern engineering enterprise will extend the colonization of the lifeworld begun in the modern era with projects and designs both "hyper-fine" and "hyper-complex."[9] The problem does not lie in any threat to health and safety, environmental sustainability, or social justice. The problem is that even the most stringently ethical hypermodern engineering enterprise will be so seductive that it will resist all critique.

Hypermodernism distorts reality in two ways. The first, more obvious way is by bringing into being what many call *hyperreality*. Hyperreality is reality as mediated through, primarily, information technologies. The hyperreal product brought into being by the hypermodern engineering enterprise tends to be seen by the typical consumer as a superior form of reality. According to Borgmann, the perception is that "it conforms more fully to the technological promise of liberation from the recalcitrance of things, the confusion of circumstances, and the foibles of human beings" and "surpasses traditional and natural reality in brilliance, richness, and pliability."[10] Its most extreme form is virtual reality simulations, which many declare can often seem more real than physical reality. The military use virtual reality simulations extensively, especially in war gaming and pilot training.

The second, more insidious way is by enhancing what Stephen Bertman calls *the power of now*. "The power of now replaces the long term with the short term, duration with immediacy, permanence with transience,

9. Borgmann, *Crossing the Postmodern Divide*, 82.
10. Ibid., 83.

memory with sensation, insight with impulse."[11] In short, hypermodernism encourages instant gratification.

Bertman enumerates some of the positive and negative effects of the power of now. "By assigning the highest priority to speed, the power of now enhances the value of those activities that exemplify speed (electronic entertainment, computerization, electronic information transfer, automation, the rapid consumption of material goods, the development of systems of belief that promise instant answers and solutions)." However, "by assigning the highest priority to speed, the power of now undermines the value of those experiences that exemplify or require time to develop (psychological maturation, the building of meaningful human relationships, the doing of careful and responsible work, the appreciation and creation of the arts, and the search for the answers to life's greatest problems and mysteries)."[12] Borgmann suggests that the only way to avoid this trap is "to outgrow technology as a way of life and put it in the service of reality, of the things that command our respect and grace our lives."[13] I would add that we can only accomplish this by aiming at *engagement with* technological processes in a thoughtful and holistic manner rather than permitting ourselves to be seduced by the lure of instant gratification and *absorbed within* them. But is short. As the hypermodern colonization of the lifeworld advances, any critique from a deeper perspective becomes less and less likely.

An interesting example of the engineering of a hypermodern reality is *government*. For the most part, in the West, the governments that now serve us are a modern phenomenon. Western democracy strives for rational, efficient, and productive government of, by, and for the people. Government, Helen Margetts tells us, is the "structure within which the activity of politics and policymaking takes place. It is the set of institutions responsible for making collective decisions for society—for solving policy problems" Margetts goes on to say that what distinguishes government is its authority—its right to rule. Following Max Weber, she explains that governmental institutions are "legal-rational" in the sense that "the institutions of executive government are characterized by rational authority, where obedience is owed to principles rather that individuals—government

11. Stephen Bertman, *Hyperculture: The Human Cost of Speed* (Westport, Conn.: Praeger Publishing, 1998), 3.

12. Ibid., 192.

13. See Borgmann, *Crossing the Postmodern Divide*, 82.

by rules. Weber saw rationality as a modernizing force, creating the logical order of premise to conclusion, consisting in linking actions and the efficient order of means to end. He foresaw bureaucracy as a phenomenon so efficient that it would spread inexorably into all areas of social life."[14]

The structure of modern government has been engineered by modern enterprises. If not countered, the structure of hypermodern electronic government will be engineered by hypermodern enterprises. The attempt to make the attainment of the common good hyperefficient and hyperproductive will entail making that effort hyperbureaucratic.

In his essay "Escaping the Iron Cage, or, Subversive Rationalization and Democratic Theory," Andrew Feenberg takes issue with Max Weber's theory of rationalization, which implies that there is an ever increasing role of the calculating and controlling aspects of social life. This is a trend that leads to what Weber called "the iron cage of bureaucracy." Feenberg insists that colonization is not a *fait accompli,* and there are ways to escape the iron cage of bureaucracy, even though as more and more dimensions of social life are structured by technologically mediated organizations such as corporations, state agencies, and medical institutions, the technological hierarchy is increasingly identified with the social and political hierarchy. Feenberg suggests that we look to the *ambivalence* inherent in technology. He insists that new technology can be used not only to preserve and reproduce social hierarchy but also "to undermine the existing social hierarchy or to force it to meet needs it had ignored. This principle explains the technical initiatives that sometimes accompany the strategies of structural reform pursued by union, environmental, and other social movements."[15] What Feenberg is talking about here are the contextualization processes stemming from the conversations of the lifeworld.

I wish I could be as optimistic. Hypermodern government will make increasing use of information technology as well as humans.[16] Even now government can move, store, transfer, link, and manipulate huge amounts of information. This is nothing less than a form of surveillance. I recall what Heidegger had to say a half a century ago about the challenging

14. Helen Margetts, "Electronic Government: Method or Madness?" (http://www.ucl.ac.uk/spp/download/publications/sppwp3.pdf), 3.

15. Andrew Feenberg, "Escaping the Iron Cage, or, Subversive Rationalization and Democratic Theory"; available at http://www-rohan.sdsu.edu/faculty/feenberg/schom1.htm/.

16. See Margetts, "Electronic Government," 3.

kind of revealing that rules in modern technology in general. He was speaking of electrical energy rather than information but the parallels are striking: "Unlocking, transforming, storing, distributing, and switching about are ways of revealing. But the revealing never simply comes to an end. Neither does it run off into the indeterminate. The revealing reveals to itself its own manifoldly interlocking paths, through regulating their course. This regulating itself is, for its part, everywhere secured. Regulating and securing even become the chief characteristics of the challenging revealing."[17]

Hypermodern electronic government will be even more capable of this challenging revealing than modern government is. Even worse, it will rationalize and systematize even more strongly the policymaking process and policy itself. It will "regulate and secure" more rigorously than merely bureaucratic modern nonelectronic government can, which is currently paving the way for it. As Margetts puts it, "The human 'machine' of Weberian bureaucracy would be delivered first by systematisation of human procedures, followed by the replacement of humans with automated machines."[18]

If uncheck by contextualization, automatization and information processing will impact and colonize all areas of social life as part and parcel of the advance of technology. Feenberg might be right to imply that hyperrationalized political reality will be more transparent and available, but I believe that we will have to become more machine-like to interact with it.

CONCLUSIONS

The modern engineering enterprise is primarily a colonizing project. Contextualization does play some role in its process, but it is out of balance. Even if the values of health and safety, environmental sustainability, and social justice, the values of process ethics, are attained, that will not be sufficient to right this balance, and if the colonization of the lifeworld by the modern engineering enterprise is not checked, there is a very real possibility that it will become irreversible.

17. Martin Heidegger, *The Question Concerning Technology and Other Essays,* trans. W. Lovitt (New York: Harper & Row, 1977), 16.

18. See Margetts, "Electronic Government," 6.

PART II

THE PREMODERN ENGINEERING ENDEAVOR

FOUR

PERSON

From the little that is known of premodern or ancient engineers, it appears that they had a lot in common with contemporary engineers. Both can be described as disciplined, dedicated, and single-minded. Hardly unexpected, since the modern engineering enterprise actually subsumed rather than supplanted the premodern engineering endeavor, inheriting its structures and functions but incorporating a scientific sensibility along with premodern methodologies. However, there are differences between premodern and modern engineers that make a difference.

For one thing, the premodern engineer tended to have a *practical* character, whereas the modern or contemporary engineer tends to be *pragmatic.* This is a crucial difference. Practical engineers used what was at hand in their world to perform useful tasks. World or context was essential to their endeavors. Pragmatic engineers, on the other hand, stress proficiency of means, independent of contexts, in the achievement of their tasks. The extent to which context grounds, informs, and conditions engineers and the engineering project is key here. For a practical premodern engineer that extent was maximal, for a pragmatic modern engineer it is minimal.

For another thing, premodern engineers tended to strive for *effectiveness,* whereas modern engineers tend to strive for *efficiency.* Effectiveness entails getting the job done, bringing forth the intended result, *within* a particular context. Efficiency entails getting the job done in an optimal fashion, ideally according to abstract mathematical measures, and ultimately without reference to context.

Premodern engineers also strove for what I call *fruitfulness,* whereas modern engineers seek maximal *productivity.* Again, these terms are often conflated. But I use fruitfulness to indicate a contextualized concern to bring into being a unique and singular entity, like a cathedral. Productivity implies a decontextualized bringing into being of a multitude of engineered entities, like the latest, smallest, fastest personal computers.

In general, then, the premodern engineer functioned in a practical, effective, fruitful manner that was necessarily grounded in contextual concerns. The modern engineer functions in a pragmatic, efficient, productive manner that tries to elude the limitations of context as much as possible.

THE PREMODERN ENGINEERING ENDEAVOR

Within the premodern engineering endeavor, the methodology was largely implicit. A premodern engineer, for example, in building a house, relied on ancient truisms like the fact that a house built on a strong foundation is a house that will endure the test of time. Such a truism can be unfolded in numerous ways. Yet only certain types and forms caught on as resonating with the spirit of those times. In modern times we call these types and forms heuristics. These forms required certain building procedures and not others, typically passed down through the years via apprenticeships that honored the traditions.

The patterns and procedures the premodern engineer employed were commonly shared and transmitted via an oral tradition. The know-how to execute a pattern was learned by practical experience. Even today there are trades and crafts whose skills are passed on through apprenticeships, formal and informal, where experience and instruction are combined in a way that would make sense to both the architects of the Parthenon and to Silicon Valley managers giving fresh engineering students some real-world hands-on training.

In the premodern era, the patterns of how things should be done were generally not formulated in a way that permitted scrutiny or interpretation. Since actual engineering practice proceeded by intuition, heuristics, rough estimates, and design experience, engineering as process, the methodology of engineering, was backgrounded in the era of traditional engineering. Unquestionably, the explicitness of the modern engineering process contrasts with the implicitness of the premodern engineering process. As an illustration of this, today a modern engineering methodology, like one for designing a low-pass electrical filter that removes high frequency signals, is readily and explicitly available in texts and handbooks.

In addition to the process, the person of the engineer in premodern times remains rather opaque. So, in fact, does the product, the engineered. Person, process, and product within the premodern engineering endeavor are all rather seen through a glass darkly. From a contemporary point of view, a look at the engineering endeavor in premodern times reveals mostly fragments of engineered products, primarily the ruins of departed civilizations. Traditional engineers of the premodern era have of course vanished, leaving only traces of their character and, here and there, their works. So, how to proceed, with process backgrounded and person and product dimly perceived? The products or works of the ancient engineers, the remains, the ruins, the fragments, have been celebrated in many fine writings on the history of engineering and technology.[1] But what about the person, the engineer? In this chapter, I will track the traces of the character of the premodern engineer and compare that character with that of the modern engineer.

One other point of interest with regard to the premodern engineering endeavor: it was not constrained or conditioned by or subsumed into a system like the modern capitalist system, which encourages productivity and efficiency and thus guides the methodology of engineering practice in the smoothest way possible toward optimal performance. It appears today that capitalism implicates engineering almost totally in its cycle of commodification, production, and consumption.

THE PREMODERN ENGINEER

The three engineering orientations I contend with in this text—premodern, modern, and focal—are only roughly affiliated with past, present, and future. The orientations, as I view them, and as I have mentioned, are subsumptive in the sense that to a large degree, the modern includes

1. I have found the following histories of technology useful: Carroll W. Pursell, *The Machine in America: A Social History of Technology* (Baltimore: Johns Hopkins University Press, 1995); Arnold Pacey's *The Maze of Ingenuity: Ideas and Idealism in the Development of Technology* (London: Allen Lane, 1974), *The Culture of Technology* (Cambridge, Mass.: MIT Press, 1983), and *Meaning in Technology* (Cambridge, Mass.: MIT Press, 1999); Klemm, *History of Western Technology;* L. Sprague de Camp, *The Ancient Engineers* (1960; reprint, New York: Ballentine Books, 1974); John Rae and Rudi Volti, *The Engineer in History* (New York: Peter Lang, 1993); and R. J. Forbes, *Man the Maker* (New York: Henry Schuman, 1950).

the premodern orientation, and the focal includes both the premodern and the modern orientations.

The premodern engineer was, and the modern engineer is, task oriented. Depending on the magnitude of the task, either the task was assigned to the engineer, or the engineer was assigned to the task. In either case, the nature of the engineered, the work or the product that was the aim of the undertaking of premodern engineering, was largely dictated to the engineer by a client or employer. This is still true today, and the contemporary engineer, like the traditional engineer, generally accepts his assignments. However, today, unlike in the era of premodern engineering, engineers can change jobs or even—in serious cases where all else fails—"blow the whistle" if unethical circumstances prevail. The premodern engineer probably did not have those options.

The premodern engineer, charged with significant and weighty tasks, surely had the freedom to exercise technical judgments, and along with freedom went certain responsibility. How the engineer exercised that freedom and carried out that responsibility is captured by the notion of *character.* But the notion of character is itself elusive, an ambiguous phenomenon that needs to be circumscribed. Carl Mitcham provides an excellent summary: "Metaphysically, character is neither essence nor accident, neither universal nor particular. Epistemologically, character is neither conceived nor perceived. Anthropologically, character is neither determination nor freedom. Character is a reality always in-between, neither universal essence nor particular accident, a limited determination and an equally limited freedom. In it lies the heart of the human condition."[2]

In spite of the ambiguities at the heart of the human condition, it does appear that engineers in ancient times had to have been, as I have mentioned, disciplined, dedicated, and serious- or single-minded. They functioned rather like modern-era foremen or officials overseeing the planning, organization, design, and construction of an engineering or architectural project. Engineer and architect were, in fact, often the same person. As John Rae and Rudi Volti in their book *The Engineer in History* maintain, in antiquity, the premodern engineer, who might be a designer and a builder of a temple, palace, fortification, harbor, road, or water supply system, was known as an *architekton* in Greece and an *architectus* in

2. Carl Mitcham, "On Character and Technology," in *Technology and the Good Life?* ed. Eric Higgs, Andrew Light, and David Strong (Chicago: University of Chicago Press, 2000), 133.

Rome.[3] The role of the ancient architect was more broadly conceived than is the role of the modern architect. Certainly today we have many different types of engineer and architect. But as an echo of a distant past, some modern structural engineers, civil engineers, have duties and responsibilities that are very much the same as the duties and responsibilities of some architects. It appears that, for the most part, the premodern and modern engineers differ more in degree than in kind.

Within their capacity as overseers of the engineering of artifacts, premodern engineers took nature as their ideal. However, they recognized, Carl Mitcham suggests, "that they could never completely duplicate the substantial union of form and matter found in nature."[4] Even though these engineers championed a synthesis of form and matter in their engineered works, utility was an essential feature of the artifacts they conceived. The premodern engineer tried to give form to matter in order to bring something into being in a *practical* contextualized manner.

We moderns downplay the synthesis of form and matter and seek, instead, a synthesis of form and function. Form serves the function, and functionality, ignoring context as much as possible, characterizes the *pragmatic* modern engineering project. The most aerodynamically designed auto, for instance, has a shape or form that minimizes drag. Today the union of form and matter, which for the ancients was intrinsic to natural things and less intrinsic but still important to artifacts, has all but vanished from modern concern. Paralleling this shift has been the shift from imitation of nature to the domination of nature that is rampant in modern culture in general, including engineering. The modern domination of nature by willful human subjectivity, in fact, aims to sever form from matter in order to convert matter to pure resource. The resource can then be shaped to serve whatever function is desired. The premodern craftsman, for example, would pick up a piece of wood and see how it inherently contained within it a wheel because of its curvature, or a fence post because of its straightness. Those forms in a way were hidden in the matter, and the craftsperson saw the forms in his "mind's eye" and proceeded to shape the matter to those forms. As another example, the stones used to build a bridge dictated to some degree the shape the bridge would take. The modern engineer takes matter as inherently formless and imposes

3. Rae and Volti, *Engineer in History*, 1.
4. Mitcham, *Thinking Through Technology*, 123.

form and function upon it. Nanotechnology is the ultimate example of this: all matter is reduced to atoms and molecules that can in principle be rearranged and assembled to any desired form and function. More traditionally, a vat of molten steel can just as easily take the form of the chassis of a Honda or a Ford, or be shaped into steel girders for building construction.

Closer to craftsmen than to applied scientists, premodern engineers were more intimate with their world, taking it more to heart and were more "oriented toward particulars instead of toward the efficient production of many things of the same kind in order to make money."[5] That is, they tended to be fruitful rather than productive. The modern engineer's movement away from particular to general concerns has allowed nature to be taken more impersonally—as a resource. The premodern grasp of reality was typically a more personal affair than we moderns are capable of or interested in, and premodern engineering was contextually enmeshed in the human world, whose "secondary qualities" the modern engineering enterprise tries to transcend as best it can. "The pre-modern or classical ontology involves looking upon matter as a living reality ordered toward taking on form—in accord with whatever form it already possesses and the potentialities contained therein."[6] Premodern reality had a keener sense of spirit as animating force than modern reality acknowledges. It's no wonder that Max Weber spoke of the "disenchantment" at the heart of the modern project.[7]

PYRAMID ENGINEERS

The pyramid is a prime example of a particular ancient engineering achievement. The design and construction of pyramids employed many engineers. What is remarkable is that "most of the truly monumental pyramids came into being in just over a century in the late Third and early Fourth Dynasties."[8] The years between 2834 and 2722 B.C. saw the

5. Ibid.

6. Ibid., 132.

7. Max Weber, *Essays in Sociology*, trans. and ed. H. H. Gerth and C. Wright Mills (New York: Oxford University Press, 1946), 155.

8. James E. McClellan III and Harold Dorn, *Science and Technology in World History* (Baltimore: Johns Hopkins University Press, 1999), 43.

construction of thirteen pyramids, including the Great Pyramid at Giza, all of which required the technical and organizational skills of the premodern engineer. The pyramids were built as tombs for the pharaohs, but pyramid building was also, according to James E. McClellan III and Harold Dorn,

> an activity pursued in its own right as an exercise in statecraft. The sequence of the early pyramids comprised giant public-works projects designed to mobilize the population during the agricultural off-season and to reinforce the idea and reality of the state in ancient Egypt. More than one pyramid arose simultaneously because a labor pool—and surely an increasingly large labor pool—was available and because the geometry of pyramids dictates that fewer laborers are required near the top of a pyramid than at the bottom, thus permitting the transfer of labor to newly started projects. Monumental building was therefore a kind of institutional muscle-flexing by the early Egyptian state, somewhat akin to the arms industry today.[9]

Farmers and slaves may have built the pyramids, but engineers engineered them. Who were these engineers? The person generally credited with the development of the pyramid was Imhotep, who was master builder for the Pharaoh Netjerikhet.[10] Imhotep engineered for the pharaoh what is considered to be the first Egyptian pyramid and the first example of natural-stone architecture, the "step" pyramid of Sakkarah around 2940 B.C. Something of the character of the ancient engineer can be seen in the epitaph another Egyptian engineer Ineni (circa 1500 B.C.) wrote for himself, as reported by Ervan Garrison: "I have become great beyond words. I will tell you about it, ye people. Listen and do the good that I did, just like me. I continued powerful in peace and met no misfortune; my years were spent in gladness. I was neither traitor nor sneak, and I did no wrong whatever. I was foreman of the foreman and I did not fail. I never hesitated but always obeyed superior order, and I never blasphemed sacred things."[11] Ineni may not have been exactly a modest fellow, but

9. Ibid., 43–44.

10. Ervan Garrison, *A History of Engineering and Technology* (New York: CRC Press, 1999), 29.

11. Rae and Volti, *Engineer in History,* 10.

he knew his place and acquiesced to his superiors. For the most part, in the existing social hierarchy, ancient engineers were comfortably ensconced between the powerful and the powerless. For example, in ancient Roman engineering, where most of the effort was directed toward various permanent structures commissioned by an emperor or prominent citizen, L. Sprague de Camp, in his classic text *The Ancient Engineers,* tells us that "a consul, senator, or other magistrate commanded the whole of such a governmental enterprise. Under him the *architectus* or engineer, in his turn, bossed a crew of minor technicians: *agrimensores* or surveyors, *libratores* or levelers, and others."[12] Most engineers today are ensconced in hierarchies in the same way the ancients, like the Romans or Ineni, were. Many today would see ideal virtues embedded in Ineni's words. The character of this engineer, or a less exaggerated version thereof, was probably typical of many premodern engineers, and is typical of contemporary engineers as well.

GREEK AND ROMAN ENGINEERS

The contextually conditioned engineering endeavor in premodern times was tied to loosely structured social and political systems, which were bound by the nondemocratic and generally repressive rule of pharaohs, emperors, patriarchs, and kings. But throughout the premodern era, at least from the Egyptian through the Greek period, the worldview contained materialistic, rationalistic, and realistic elements, but it also contained sacred *memes,* and a certain take on these memes formed the parlance of the premodern engineer.

Through all the vicissitudes of the premodern era, engineers proceeded with raising their city walls and building their temples and palaces. They paved their roads, erected their aqueducts, dug their canals, tinkered with their machines, and soberly built upon the discoveries of those who had gone before them.[13] These engineers—innovators, architects, designers, builders, inventors—were a rare breed. But being the rare breed often makes the engineer the strange bird, the odd duck. Part of the reason for this is that engineers have always been both doers and thinkers, and that tended to be at odds with the given reality wherein people were either

12. De Camp, *Ancient Engineers,* 165.
13. Ibid., 26.

one or the other but not both. The status of the engineer was below that of the priest but above that of the ordinary citizen. The ruling class, who along with the priests constituted the upper echelons of society, fancied themselves as cultured and philosophically minded, certainly not your garden-variety "hands-on" types. The elites, especially in the Greek period, and especially the intellectual class, didn't seem to care too much for the engineer but they did value what the engineer could do for them. That attitude persists today in that the upper class tends to look down its collective nose at the middle and working classes, and engineers who are upwardly mobile tend to move into middle-class circumstances from working-class roots. There is the famous story about Herbert Hoover, engineer and U.S. president, who made the acquaintance of a lady on a steamship:

> Tell me, Mr. Hoover, what are your interests?
> Madam, I am an Engineer.
> Really? I took you for a gentleman.[14]

The inegalitarian Greek society was rather indifferent to its engineers, who were unnoticed by the ordinary citizen and ignored by the elites. After democracy finally did appear in Athens, its existence apparently had little effect on engineering per se. The Greeks were never the engineers that the Egyptians had been or the Romans turned out to be. But Greek engineers, using five simple machines, namely, lever, wheel, pulley, wedge, and screw[15] did bring into being spectacular works, like the Colossus of Rhodes and the Pharos or Lighthouse of Alexandria, and, of course, the Parthenon.[16]

Reasons the Greeks never promoted a vigorous engagement with engineering, in spite of their obvious successes, can perhaps be found by considering the Greek mind. Among the educated Greeks could be found many lovers of wisdom, who valued being and thinking over doing and making. They found practical engagements unsuitable and did not seem to be inclined toward engineering.

Within Greek thinking, several distinctions took shape for the first time in a lucid and distinct fashion. Chief among these was the distinction between theory and practice. Out of philosophical stress on theory

14. See http://www.cen.bris.ac.uk/resource/engquot2.htm/.
15. Garrison, *History of Engineering and Technology*, 51.
16. Ibid., 45.

and the supersensuous Realm of Ideas emerged the great cultural achievement of ancient Greece, the development of a scientific sense.[17] But their science was not like ours. Ancient science was based on observation and classification. It had little interest in experiment. The ancients studied nature to understand it, not to conquer it for the sake of the betterment of the human condition. R. J. Forbes even claimed that "generally speaking Greek scientists had a horror of manual work and despised those who had to engage in it. They loved theory, but when discussing how to use this knowledge Aristotle and Plato are definite in their rejection of practical applied science as a proper task for the scientist."[18]

Roman engineering was much more heralded than Greek engineering. The Greek was the theorist, but the Roman was the doer. Roman engineers were primarily builders. They constructed roads, harbor works, aqueducts, temples, forums, town halls, arenas, baths, and sewers.[19] In engineering their many wondrous things, the Romans had little use for science. The abstract realms of theoretical deliberation lacked an immediate utility that could be brought to bear on the construction of projects such as the extensive system of Roman roads and aqueducts that consolidated the infrastructure of the Roman Republic and, later, the Roman Empire. A concrete substance, like cement for instance, took precedence over an abstraction like, say, the idea of the place of cement in a hierarchy of building materials. As McClellan and Dorn put it: "Less lofty perhaps, but no less important as a building block of Roman civilization, the invention of cement was a key new technology introduced by the Romans, one that made stone construction much cheaper and easier, and it literally cemented the expansion of the Roman empire."[20]

Among the Romans, then, there was very little appreciation of having theoretical science augment the practical engineering endeavor. Yet Forbes insists it is only a "part-truth" that the Roman engineer despised science. He (seldom she) despised *theoretical* science but not *applied* science: "He observed nature but was a poor biologist; he was a good engineer but a mediocre mathematician; he reformed the calendar but made no significant contributions to astronomy. . . . Science was valued only so far as it had practical and useful results for the State. The Roman did not have

17. Klemm, *History of Western Technology*, 18–19.
18. Forbes, *Man the Maker*, 62.
19. De Camp, *Ancient Engineers*, 164.
20. McClellan and Dorn, *Science and Technology in World History*, 89.

the patience of the Hellenistic scientist, who sought to discover the laws of nature."[21]

In the first century B.C., the Roman engineer and architect Marcus Vitruvius Pollio wrote a book *De Architectura* in which he proposed that the appropriate aim of the builder's efforts was construction of structures that had "firmness, commodity, and delight" or "strength, utility, and aesthetic value."[22] According to Vitruvius, the ideal engineer (*architectus*) was "a man of letters, a skilled draftsman, a mathematician, familiar with historical studies, a diligent student of philosophy, acquainted with music; not ignorant of medicine, learned in the responses of jurisconsults, familiar with astronomy and astronomical calculations."[23]

Although such a paragon probably never existed, Vitruvius was painting a picture of an ideal type. We try to educate our modern engineers to incorporate a specialized discipline, within a broad-based holistic perspective. Although the premodern engineer was more of a generalist who was adept at performing a large number of diverse tasks, often simultaneously[24]—and the modern engineer functions more like a specialized professional applied scientist, what they have in common is that they both bring things into being, the premodern engineer in a practical way, the modern engineer in a more pragmatic way. They bring into the flux of our worlds products that can disburden us or enliven our lives or at least entertain us. Is there a general portrayal we can put forth for these bringers into being? Whatever is brought into being is done so by the engineer for the benefit of humans, if only for the benefit of the person whose idea prevailed in the market place of ideas. The engineer is a benefactor. From the dustiest ancient engineer to the brightest postmodern, the promise or the *telos* of the project of engineering has been, is, and will be the improvement of the human condition.

CHARACTER ISSUES

In his famous essay "The Question Concerning Technology," Martin Heidegger distinguishes the forester of today from his grandfather who lived

21. Forbes, *Man the Maker,* 69–70.

22. From the translation of Vitruvius's work available at http://www.ukans.edu/history/index/europe/ancient_rome/E/Roman/Texts/Vitruvius/1.html/.

23. De Camp, *Ancient Engineers,* 165.

24. Claude Lévi-Strauss, *The Savage Mind* (Chicago: University of Chicago Press, 1962), 17.

in a more premodern manner. The modern forester "engineers" the forest. The grandfather forester was more in tune with the forest, ministering to it in a caring way. He and it were one. The modern era grandson takes the forest as a giant resource, which is integrated with modern technological systems, which in turn serve the elements of the capitalist system. "The forester who, in the wood, measures the felled timber and to all appearances walks the same forest path in the same way as did his grandfather is today commanded by profit-making in the lumber industry, whether he knows it or not. He is made subordinate to the orderability of cellulose, which for its part is challenged forth by the need for paper, which is then delivered to newspapers and illustrated magazines."[25] I would argue that both the grandfather and the grandson have caring and pragmatic or practical characters, but with the grandfather being more toward the caring end of the spectrum and the grandson, in the grip of the "grab"[26] of the capitalist system, more toward the pragmatic end.

That character issues should be paramount in an investigation into the person of the engineer leads us to look at the work of Samuel Smiles, one of the first people to write about the character of the engineer. Smiles maintained that the successful eighteenth- or nineteenth-century engineer was "orderly, regular in his habits, disciplined, predictable, methodical in his problem solving, even-tempered, and law-abiding."[27] A straight shooter. Not cynical like many moderns and postmoderns. The virtues Smiles pointed out would benefit the premodern engineer of ancient times and the early modern engineer of the eighteenth and nineteenth centuries. Many of the virtues encouraged in the character of yesterday's engineer would support today's engineer and tomorrow's engineer as well.

Smiles's books were very popular in the nineteenth century. They mirrored the individualism that was advancing in the modern era. The rugged individual was producing more and more for consumers, commodious individuals, to consume.[28] And these individuals, as Borgmann maintained and as I mentioned in an earlier chapter, were often the same person.

25. Martin Heidegger, "The Question Concerning Technology," in *The Question Concerning Technology and Other Essays,* trans. William Lovitt (New York: Harper & Row, 1977), 18.

26. Michael Eldred, "Capital and Technology: Marx and Heidegger"; available at http://www.webcom.com/artefact/artfinvn.html/.

27. T. P. Hughes, "Introduction," in Samuel Smiles, *Selections from the Lives of the Engineers* (Cambridge, Mass.: MIT Press, 1966), 11.

28. Borgmann, *Crossing the Postmodern Divide,* 38.

Individuals of high principle and integrity, who were honest, open-minded, and industrious—as championed by Smiles—could be entrusted to bring forth a world worth living in. By writing about engineers, inventors, and industrialists as they transformed their environment—and society—through rapid industrialization, Smiles not only reflected his age, but also influenced it.[29]

In his *Lives of the Engineers,* Smiles tells the story of several engineers, including James Brindley, John Rennie, and Thomas Telford. Brindley was an interesting example of what I am calling a premodern engineer, even though he lived in the early modern era. He was a self-taught genius. He could only minimally read and write. Yet he was very observant and

> ready at devising the best methods of overcoming material difficulties, and possessed of a powerful and correct judgment in matters of business. Where any emergency arose, his quick invention and ingenuity, cultivated by experience, enabled him almost at once unerringly to suggest the best means of providing for it. His ability in this way was so remarkable, that those about him attributed the process by which he arrived at his conclusions rather to instinct than reflection—the true instinct of genius.[30]

The lack of a modern scientific method or procedure did not stop Brindley or engineers of the premodern era from the enactment of monumental projects and the achievement of great works. Intuition, instinct, and experience—pivotal to the practical skill set and know-how of premodern engineers—were revealed, not in any kind of systematic manner, but rather via their character, via the ways they conducted their engineering life. The power of character, so it appears, compensated the premodern engineer for the lack of explicit methods, means, and procedures. Character is developed over a long period of time and requires the practice of the virtues. And what, again, is character? Character is a power, a faculty, the way one is, one's status or capacity. Or as Ralph Waldo Emerson put it, "This is what we call Character—a reserved force that acts directly by presence, without means."[31]

29. See Hughes, "Introduction," 1.

30. Samuel Smiles, *Selections from the Lives of the Engineers* (Cambridge, Mass.: MIT Press, 1966), 166.

31. Ralph Waldo Emerson, *The Essays of Ralph Waldo Emerson* (New York: Random House, 1944), 270.

The presence of the character of the engineer exhibits continuity. Recall that, not too long ago, engineers used to stand out in a crowd as nerdy white males with crew-cuts, plastic pocket protectors, and slide rules on their belts. Today's engineer is more hip, and he or she is likely to be non-white. In fact, in the United States, diversity is becoming more pronounced in engineering than in most other professions. Yet the character of the typical engineer is much the same as it ever was. Characteristics this character exhibits include, as I have indicated, discipline, dedication, persistence, and patience. In their everyday involvements engineers carry out a plethora of functions, like research, design, development, production, operation, maintenance, sales, and management. The engineer enacts the role of engineer, and character is manifest in the ways this role is enacted. Because being an engineer requires focus and discipline, the character of the engineer cannot be easily developed or transformed. It cannot be donned or doffed like a "persona." The social role of the *engineer as person* certainly contributes to her or his character as the *person as engineer* embraces the contingencies and contexts of professional life. Role shapes character but character orients one toward role. And roles are enacted in the human lifeworld.

As an example, several times I played the role of academic ambassador for my college of engineering by participating in our Outreach Program. In venturing out of the hallowed halls of engineering education into the *real* world to talk to high school students, I would for instance say something like: if you want to consider a career in engineering, you should be turned on by math and science and accustomed to sizing up problem situations and visualizing a solution. Being a problem solver is of the essence. Creativity and imagination as well as an analytical mind are paramount. You must also be practiced at the arts of ethical judgment and human communication. All of these attributes, plus the customs and practices that go with them, tend to shape a homogeneous character in the enactment of the role of a typical engineer, in the person as engineer. This is in spite of the wide heterogeneity in styles of the engineer as person, especially in contemporary times, but also in the preindustrial era.

Character is often associated with the notion of *ethos*. Ethos can be viewed as the habits or attitudes definitive of a particular group, the fundamental values or spirit of the group.[32] The group we have in mind is

32. Carl Mitcham, "Computer Ethos, Computer Ethics," *Research in Philosophy and Technology* 8 (1985): 276.

the collection of individuals called engineers. There is another sense of ethos, more broadly construed, as the *place* wherein character is revealed. Habits develop within habitats. Ethos as habit and habitat: the ethos of engineering as both habits or character of engineers as well as habitats of the engineering project. The habitat refers to the social and historical contexts within which engineers engineer. Premodern engineers in-dwell with that habitat, while modern engineers ignore it as best they can, and focal engineers reappropriate habitat as intrinsic to their engineering practice.

HEIDEGGER'S TAKE

To integrate the character of engineering with the notion of ethos as place, I invoke the Heideggerian idea of the person of the engineer as *Dasein*. Dasein literally means being-there. It refers to everyday human existence, which always occurs in some specific manner, in some specific place. Dasein *takes* place, by adopting various involvements in the world, but Dasein also *gives* way, by letting its world unfold in the way it does. As well as being a socially active adventure, human life is an engaged kind of letting, a releasement. When I think about this, I think of the sport of curling, wherein a stone is pushed on the ice and one object of the game is to clear the path, to make way for the stone.

As Heidegger says: "Releasement toward things and openness to the mystery belong together. They grant us the possibility of dwelling in the world in a totally different way. They promise us a new ground and foundation upon which we can stand and endure in the world of technology without being imperiled by it."[33] With the right frame of mind, the engineer as Dasein can stand and endure, stand and deliver, stand up and be counted, among a variety of ways to be. The impression I get from the Heideggerian view of all this is that technology has a certain *gravitas*. The pull of the earth, not to be denied, is the voice of context, seeking to be acknowledged. Technology is a serious endeavor, not something to be taken lightly, because technology pervades and conditions all of reality—particularly modern engineering reality. Ironically, it is modern reality that tries as best it can to evade the *gravitas* of technology. The engineer,

33. Martin Heidegger, *Discourse on Thinking*, trans. John M. Anderson and E. Hans Freund (New York: Harper & Row, 1966), 55.

at least, must remain vigilant, especially with regard to the long term effects of technology, and must direct the power of technology in a caring manner in order to bring into being the fruits of technology's beneficence, the promise of an authentically engineered reality.

Another implication of the Heideggerian take on human existence is that human existence is to be construed in a holistic manner. In other words, humans find themselves always already immersed in a world, thrown into a world teeming with life, with other humans, and with the things we need to keep our lives alive. World is not somehow added onto the notion of person but it is the concrete "wherein" within which we find ourselves. More abstract notions of self as subject, as ego, as consciousness, or as mind in a body/mind dualism are available as detachments from this primal condition of being-in-the-world. Fundamental to Heidegger's notion of Dasein is that human existence consists of three co-equal radical structures which he called *projection, disposition,* and *fallenness.* Philosopher John Caputo elaborates: in the sense that we humans are "projected," we are always ahead of ourselves, cast forth into one course of action or another. In the sense that we are "disposed," we are cast forth or thrown into the world and already situated within pregiven circumstances. In the sense that we are "fallen," we are the being which projects himself ahead, from out of a given situation, ever liable to give up his project and to sink back into complacency with everyday reality.[34]

Typical engineers, then, would find themselves engaged in a course of action called engineering. They would be conditioned by the circumstances of what it means to be an engineer. The traditions, values, cultures, institutions, meanings, and legacies of the engineering project are all part of this conditioning process. Then, of course, there is always the possibility of fallenness, turning away from disposition and projection. In fact, such turning away is really the more typical stance, because when one is doing engineering, one must be focused, for example, on the set of equations that need to be solved, right here and now, as a step in this or that engineering process. Doing engineering and reflecting on that doing are two different things.

In a further development of the Dasein idea, Heidegger related disposition to the past, projection to the future, and fallenness to the present.

34. John Caputo, "A Phenomenology of Moral Sensibility: Moral Emotion," 3; available at http://www.crvp.org/book/Series01/I-12/chapter_ii.html/.

We are oriented toward the future via the way we understand things, and we understand by projecting—to some degree having already projected—our possibilities onto the things in our world. But things in our world have always already projected their possibilities onto us in their disposition toward us, and in discovering this we find ourselves *disposed,* that is to say, in a mood or with a feeling toward the world, an affectivity. We discover ourselves, then, with an affectivity and an understanding which reveal to us how we are in the world. All this occurs in what Heidegger calls an "equi-primordial pre-thematic fashion." From that grounding, we can begin to refine our understanding by virtue of philosophical interpretations of various sorts. That refinement can induce shifts in affectivity, which can further tune our understandings. There is an ongoing backward and forward relatedness here, as we get closer and closer to true self realization or disclosure or what Heidegger would call *authenticity.* Of course, as I have mentioned, standing in the wings of the here and now is the ever present tendency toward fallenness which is the tendency to give up on the project of self-disclosure, to let the circumstances of our lives dictate how we will be. I can cease and desist from recollecting the "thrown" circumstances of my life and the affectivity that goes with them, and I can give up on the strenuous work of sorting and sifting the projection of my possibilities I had labored so hard to organize. I can take my cue from television, for instance. Do what they do, try to look like them, see what they see, buy what they buy. Or, as mentioned earlier, circumstances might demand, and often do demand, that I bracket any concern with looking at how I am, and require me to get on with some task that is at hand. This is true of both the premodern and modern engineer.

Most certainly we slide in and out of fallenness all the time. In the Catholic tradition there is a term *concupiscence,* which means a tendency to sin, to fall away from grace. It is indicative of the human condition and follows from the sin of Adam, original sin. Heidegger, who was raised a Catholic, probably had something like this in mind in his term "fallenness," although he insisted that fallenness had nothing to do with sin or morality. It had to do with the examined life and the unexamined life. The latter may be not worth living, but it is not a sin. And when we fall into this unexamined-ness, do we actually fall, or do we leap, or are we pushed? A friend insists that he is "bound, taken, thrown, fallen, and abandoned." The question here is: are we just victims or do we have some say in the matter? Did the devil make me do it? Or there is the old saying that falling down is part of growing up. In any event, as far as character is concerned, the way I am as an engineer is revealed in the character I exhibit in the actions I perform. And that is a concupiscent character, no doubt fragmented and tattered to various degrees.

The world in which all this transpires is my ethos as place. But ethos as character, exhibited in the practice of the virtues, is more than passive habit. I may be disposed in certain ways in a passive sort of manner, but that disposition goes with a projection, an action, whereby world as place is opened up. I have to open up to the world in order that the world can show itself to me. And I can only take what is given. In moments of vision, when I am not doing engineering but rather reflecting on what it means to be an engineer, the affectivity I am disposed toward tunes my projected understandings as my understandings tune my affectivity. What can be said about the attunements of the typical premodern engineer?

THE PREMODERN ENGINEER'S ATTUNEMENT

"The machine unmakes the man. Now that the machine is so perfect, the engineer is nobody."[35] Thus spake Ralph Waldo Emerson in the mid-nineteenth century. As we enter the twenty-first century, the machine has become even more perfect. The engineer then should have all but disappeared, if Emerson had it right. But today the engineer is everywhere. In fact, the engineering metaphor has become ubiquitous. We engineer our values, our financial portfolios, our relationships. The nobody Emerson spoke of has today become the everybody. The character and attunements of the typical engineer, then, can perhaps be found in looking at "everyman" who is all of us as we go about the day to day work of engineering our lives and our worlds. But here the notion of engineering has been reduced to the notion of making. The usurpation by "everyman" of the character of the engineer is symptomatic of the notion of fallenness, which Heidegger claimed was intrinsic to the existential structures characterizing Dasein. To bring the character of the premodern engineer into view in terms of understandings and affectivity will require a certain resistance to ready-made interpretations.

35. This quotation comes from Ralph Waldo Emerson, *Society and Solitude;* see http://www.scottj.f9.co.uk/engquot.html/.

Surely engineers, even the premodern ones, are more than just makers. They are designers, testers, manufacturers, maintainers, and so on, of systems, devices, organisms, networks, and structures. The attributes of character of premodern engineers that we have so far unearthed, in addition to discipline, dedication, and single-mindedness (or stubbornness), include sobriety, rationality, and courage, as well as a willfulness involving enthusiasm and energy. As Smiles said, he or she is orderly, has regular habits, is predictable, methodical about problem-solving, even-tempered, and law-abiding. Engineers are honest, open-minded, and industrious. They are solid, steady, persistent, and patient. Engineers have creativity, imagination, sensible ethical judgment, and good communication skills. Most engineers would some of the time exhibit some of these ways of being, knowing, and doing. Unlike the modern engineer, who tends to be practical, efficient, and productive, the premodern engineer tended to be pragmatic, effective, and fruitful.

Who, then and again, is this person, this engineer? According to the famous engineer James Kip Finch: "The engineer has been, and is, a maker of history."[36] And, I would maintain, a maker of futures as well. To make history and futures, the engineer needs to contribute to the bringing into being of the things that matter to our cultures and civilizations. As benefactors of humankind, engineers need to have an attunement of their affectivity/understanding (or as Heidegger might say, *Befindlichkeit/Verstehen*) constellation that yields at least a disciplined and dedicated character, even though that character can play itself out in several ways.

THE MEDIEVAL ENGINEER

We have an image, then, of who the engineer was and is. The image of the ancient engineer is fairly consistent with that of the modern engineer, except for the stress on context that informed the ancient engineer and the decontextualized nature of modern engineering. Between ancient and modern times, however, lies the medieval period. From the available evidence, there too, at least in Western Europe, the character of the engineer pretty much stayed the same. The person of the engineer, his or her temperament, held rather steady in the face of the shifting winds of the prevailing medieval *Zeitgeist*. The period from around A.D. 500 to around

36. James Kip Finch; see http://www.asme.org/history/hquote1.html/.

A.D. 1400 is typically viewed as rather chaotic. The old Roman system of a strong central government gave way to a multitude of independent and self-sufficient towns, duchies, and fiefdoms. R. J. Forbes maintained that the collapse of "central authority had profound effects on the development of engineering, for only a central authority is strong enough to organize and finance expensive public works like roads building, canals, and bridges."[37] Engineering became a local and decentralized endeavor. It contracted. But this was only during the early medieval period. The tide began to turn, local endeavors began to expand into larger enterprises, around the turn of the millennium.

In fact, from ancient to contemporary times, there have always been modulations at the heart of the human condition. The materialist versus antimaterialist cultural distinction is a case in point. Early on in ancient times, according to Taichi Sakaiya, there was great potential for vastly expanding the amount of arable land. Surplus production gave birth to a system that needed and utilized human beings as "tools with motive power." Slavery that had existed on a very limited scale began to be implemented on a grander scale. As a result, more and more goods were produced, and more and more commerce could thus be carried out.[38] In ancient times a high level of material culture emerged—at least for the average citizen—and held sway for hundreds of years, until around the time of Christ and the rise of an antimaterialist ideology.

For the average person, however, the new abundance in ancient societies had resulted in a materially richer world. We moderns enjoy a similar abundance, and in this way materialism characterizes both modern and ancient cultures. Both modern and ancient societies exhibit notions of taste and ethical systems that define happiness in terms of a bountiful supply of goods, and justice as anything that contributes to such bounty.[39]

But the ancient world gave way to forces of overpopulation and resource-depletion, as well as economic corruption. More pointedly, however, the Western Roman Empire was destroyed by Germanic invaders from the north and the Western Jin dynasty in China was destroyed by the attacking nomads from the steppes along the northern boarders of China. Similar attacks destroyed the Gupta dynasty that ruled India. But, according to Sakaiya, as mentioned earlier, these attacks were not the

37. Forbes, *Man the Maker,* 103–4.
38. Sakaiya, *Knowledge-Value Revolution,* 159.
39. Ibid., 134.

real cause of the decline in these civilizations. "It was the change in tastes and ethics that had been under way long before that."[40] And these changes, in turn, were connected to the rising scarcity of resources. The shift was from the materialistic values of the past and toward a more spiritual kind of culture. It was a shift toward a society with less material consumption and objectivity and more of a nonrational spirit that embraced a kind of social subjectivity and antimaterialism. With the disappearance of resources to make goods to consume, people with social conscience began to protest materialism. Consumption, it turned out, was not the only path to happiness. In fact, it was the wrong path, in light of the antimaterialistic spirit that was on the rise.

The Middle Ages, whether in Europe, India, China, or the Middle East, could be characterized, according to Sakaiya as "a lack of goods, a surfeit of time."[41] The time was devoted to contemplation, meditation, and prayer, or to leisure activity. This earthly vale of tears, particularly in the Western Christian view, was considered nothing more than a prelude to an afterlife in a glorious heavenly abode. The paradigmatic figures of the Western European Middle Ages were people like St. Francis of Assisi, who walked a spiritual path, practicing love, honesty, poverty, and abstinence. He lived a life untouched by economic concerns.

As indicated earlier and as might be expected, scientific and engineering advances during this period were somewhat stagnant, especially early on in the Western medieval era, largely caused by a decline in central authority. Engineers without monumental projects to engage them tended to become craftspeople. Yet there were several achievements of the medieval period that did engage the premodern engineer, and people of the medieval period did eventually begin to take an active interest in promoting the engineering of useful things, once it became apparent that many devices provided an effective work-saving service to society. Though productivity itself was never explicitly pursued, the practical benefits of it started to become more and more apparent in the engineering of a better kind of human life. Since the need or desire to be productive in any systematic way only emerged very slowly, however, so too did any developments in science and engineering.

What kinds of things were engineered by the premodern engineer in

40. Ibid., 162.
41. Ibid., 177.

the Middle Ages? Cathedrals and mosques were places of prayer and worship that were among the most spectacular of engineered structures. But they were few and far between. On a more down to earth level, the steel plow, introduced around A.D. 600, became a boon to agriculture, especially in Northern Europe where the soil was rich and loamy. Medieval engineers utilized the elemental forces of beast, water, and wind to a far greater degree than was possible in antiquity, where slave power was the primary force. Although the use of wind power by means of a wind-driven wheel was practically unknown in antiquity, there was the design of Hero of Alexandria in the first century A.D. to utilize wind to drive an organ pump. As Klemm tells us: "The medieval change-over to the application of natural sources of power betokened technical progress which had improved results, comparable in modern times only to those following the introduction of the steam engine in the eighteenth century and the utilization of atomic energy in our own day."[42] (This last quotation by Friedrich Klemm, it should be pointed out, comes from his book which was first published in 1954 when the Atomic Age was in full-bloom. Today we are not so sanguine about the prospects of atomic energy.)

The engineers of antiquity played with the forces of steam and wind, and actually were able to design and build machinery that worked by moving weights or by air pressure, and sometimes even by heat.[43] But there was little incentive for ancient engineers to harness these forces. If the development and production of these devices had proceeded in earnest, what kind of world would have come about? Would the industrial revolution have occurred a thousand years earlier?

Windmills that were more than mere playthings, that actually produced power for useful purposes, were invented, probably sometime in the early seventh century A.D., in the region called Seistan, a border lowland region of southwestern Afghanistan and eastern Iran. Seistan was very dry and could not depend on running water for power, but it was quite windy. Windmills were the ideal source of power. The mills were supported on substructures that provided elevation or on the towers of castles or on the tops of hills.[44] That region was in those days part of the Persian empire. Legend has it that the slave Abu Lulua constructed the

42. Klemm, *History of Western Technology*, 79–80.

43. Forbes, *Man the Maker*, 62–63.

44. Donald Routledge Hill, *Engineering in Classical and Medieval Times* (London: Routledge, 1996), 103.

first windmill for the Caliph Omar.[45] Unlike the later European windmills with horizontal windshafts, the Persian windmill had vertical windshafts with sails. This type of windmill, over the next few hundred years, spread throughout the Islamic world and eventually to China and India. In Egypt in the Middle Ages it was used in the sugarcane industry, but its most extensive application was to gristmilling.[46] It was not until late in the twelfth century that windmills appeared in Europe, the first one being built in Normandy from where its use spread to the Low Countries and across the channel to England. It soon became the power plant of choice, especially along the flat wind-blown regions of the European Atlantic seacoast.

The premodern engineer of the medieval period who engaged in the design and construction of windmills, waterwheels, and agricultural technologies of various sorts must have been careful, patient, disciplined, and dedicated to the tasks at hand. A person grounded in context. The patterns of windmill design, for example, were probably passed along to the engineer from elders and mentors who kept alive a tradition of useful device development. In addition, many technologies first developed in the Middle East were brought to Western Europe by travelers and crusaders.

CONCLUSIONS

I was attempting in this chapter to lay out the elements of the character of the engineer, especially the character of the premodern engineer. The person as engineer was my chief concern. The premodern and the modern engineer share much common ground. But the business of articulating a fixed set of features of the engineer seems to keep unfolding into more and more "types." Samuel Florman insists that we acknowledge the variety of types of people who become engineers, and that sweeping generalizations are dangerous. "From the 'rank and file' to the most productive creators, from the angry and alienated to the inspired and committed, the profession contains a wide variety of human types—as well it should, engineering being an elemental expression of the human spirit."[47] Yet we have some character traits that must be put forward as ideals. What Florman is talking

45. Forbes, *Man the Maker,* 93–94.
46. Hill, *Engineering in Classical and Medieval Times,* 103.
47. Samuel C. Florman, *The Introspective Engineer* (New York: St. Martin's Griffin, 1996), 125.

about is the *engineer as person,* that is to say, in terms of personality types; and I am summarizing the *person as engineer,* namely, in terms of character traits. The engineer as person opens up the discussion to the huge variety of ways of being human. The person as engineer focuses the discussion onto those particular traits that constitute the essence of the engineer.

Just as designing an attitude-control system, a typical task for an aerospace engineer with a control theory background, requires a delicate balance of parameters, so too the articulation of the attitude or character of the person of the engineer requires some delicate balancing as well. The person of the engineer, balanced and harmonized in the ideal case, is disciplined, dedicated, serious, literal (but not inflexible), industrious, rational, persistent, curious, patient, creative, and imaginative. Those attributes would seem to serve the traditional engineer as well as the modern engineer. The modern engineer, according to Ted Hissey, a former director of the IEEE, also needs a global perspective, must be a team player, and must have multiplexing capabilities.[48] Of course, these days the more computer skills an engineer has the better off he or she will be.

Thus, though the skill sets change, the character of the engineer has held rather steady over the past several millennia. These character traits really are appropriate for any kind of professional person. Should not the lawyer be rational and the physician creative? One major difference between the engineer and members of other learned and skilled professions is that we engineers focus primarily on artifacts, and only secondarily and indirectly on humans themselves. However, that indirect and secondary reference to humans is precisely where the humanizing of the engineering project has its point of departure. Recognizing that the products of our endeavors are for the sake of other humans can be the first step in tuning us to the collegiality we all share as professionals.

48. T. W. (Ted) Hissey, "Enhanced Skills for Engineers," *IEEE Grid* (a newsletter of the IEEE), February 2001.

FIVE

VIRTUE ETHICS

In the previous chapter, I investigated the character of the premodern engineer, who came into being and flourished in the time period roughly from the pyramids to the cathedrals. The spirit of that character, even today, is very much present within engineering practice, especially in those projects which employ heuristic and intuitive methods. The character of the premodern engineer, in fact, was seen to have much in common with the character of the modern or contemporary engineer. The character of the premodern engineer was and is gauged by ethical and moral assessments. The primary ethics extant in the premodern era was virtue ethics, and that was and is the type of ethics appropriate to the premodern engineer.

The culture of engineering is shaped by the practice and conduct of individual engineers who adhere to certain moral and technical values, for example, thoroughness, honesty, industriousness, and efficiency. Those values are manifest in the character of engineers at work on a given engineering project. In the previous chapter, I presented a general view of the character of the premodern engineer at work within the culture of the premodern engineering endeavor. Character, habits, and customs generally become ingrained in an unreflective manner. For the most part, though not entirely or necessarily, we adopt them or adapt to them with minimal deliberation. According to Gregory Trianosky, "Character is the product not only of voluntary action but also of the activity of temperament, along with upbringing, childhood experiences, social environment, peer expectations, and pure happenstance."[1]

In general, habits may be deemed virtuous, as John Dewey maintains,

1. Gregory Trianosky, "Natural Affection and Responsibility for Character: A Critique of Kantian Views of the Virtues," in *Identity, Character, and Morality: Essays in Moral Psychology,* ed. by O. Flanagan and A. O. Rorty (Cambridge, Mass.: MIT Press, 1990), 104.

because they are sustained by social approval and admiration.[2] But such virtuous habits constitute what Dewey calls *customary* morality as distinct from *reflective* morality. Customary morality is pre-reflective. The customary morality of the typical premodern or modern engineer is exhibited in the array of her or his character traits. This is the way we have come to be as a result of social approval for the things we do. In this chapter, I turn toward reflective morality, and look at development of a system of virtue ethics that might have supported the moral life of premodern engineers. Plato insisted that basic values were the foundation of moral and immoral behavior which may exist in any organized whole. As Sherwin Klein puts it, "The values that guide those wholes mold the character of people who participate in the organizations, and the moral and immoral conduct exhibited in organizations is a product of the character of its people."[3]

A STORY

In 1968, I landed my first job in the booming Los Angeles aerospace industry. I was just out of college with an M.S. degree in electrical engineering, proud to be contributing to the American effort to get a man on the moon. But by 1970, aerospace work began to lose its luster for me. The moon landing in 1969 was a peak experience for the space program, and after that things began to flatten out. So I decided to return to graduate school to study electrical engineering and also to try to "find myself." I didn't. But I did get a Ph.D. in electrical engineering and in the mid-1970s set out in pursuit of an academic position. They were few and far between. Positions in general in engineering and science were not plentiful in those turbulent and economically recessive times of the mid-1970s, characterized by the winding down of the Vietnam War and a peaking of opposition to the military-industrial complex.

While finishing graduate work in engineering and completing a thesis in the area of differential game theory, I was simultaneously attempting to give myself a liberal arts education, which I had successfully avoided

2. John Dewey, *The Theory of the Moral Life* (New York: Holt, Rinehart and Winston, 1960), 91.

3. Sherwin Klein, "Platonic Virtue Theory and Business Ethics," *Business and Professional Ethics Journal* 8, no. 4 (1990): 65.

as an undergraduate. I did succeed in deepening my interest in philosophy, politics, and poetry and was looking for some way to apply the technical work I had done to a domain of discourse that promoted the social good in a tasteful if not poetic sense. Though there was poetry, I thought, in some of the beautiful equations of differential game theory, poetry and game theory, by and large, seemed miles apart. Most of the work in differential game theory, in fact, was in the arena of military applications and war games. So my quest for an ideal career was going nowhere; however, some industrial positions were becoming available, and after much job searching as well as soul searching (and many hours working the register at a convenience store), I landed an interview with a large aerospace company in the Silicon Valley. I was delighted to finally be having an interview. During the interview the engineers in my prospective group enthusiastically described the system they were developing. It sensed the terrain with an ingenious radar mechanism, employed an elaborate feedback control structure, and made determinations on the basis of statistical decision rules. The job offered fascinating prospects for sophisticated engineering designs. But after the interview I perused the material I had been given and took a wider look at the project. The technical details were so interesting to me at the time of the interview that I failed to realize the system I'd be working on was to form part of the signal processing unit of the "aircraft" that came to be the cruise missile.

Here was a serious dilemma. I had come to think that war was good for "absolutely nothing" except possibly making the rich people richer. The question I had to face was, Could I contribute to such a project and thereby to the militarization of the planet? This is not an uncommon problem in technoscience fields: here was a technically "sweet" project whose intended use was at best unsettling. What to do? My instincts told me to walk away. I did and was fortunate soon afterward to get an academic position, which was more suitable to my temperament.

Reflecting on this situation, I realized that I had not explicitly employed any particular type of engineering ethics, any system of ethical thought, to guide my decision about a situation that had obvious moral import. I didn't, for example, make the decision by seeking out the greatest good for the greatest number. I came to a decision based on my character, my internalized values or *ethos*, my customary morality. The problem is that this scenario wherein personal values rather than moral systems guide ethical decisions appears to be quite common, as I mentioned earlier. Such an approach might be adequate for ordinary dilemmas, but with complex dilemmas, often a more explicit approach is called for. The question arises, What values should engineers and engineering cultivate in an explicit manner?

Even though systems of ethical thought grounded in abstract principles began to appear shortly after the dawn of the modern age, engineering ethics did not come into its own as a separate discipline until the late 1970s, focusing its concerns on codes of ethics. Codes existed prior to the 1970s—in fact, since the late 1800s for most traditional fields of engineering—but engineering ethics in general and codes in particular were relatively unknown to most, including me at the time I faced my ethical dilemma. Codes did attempt to encapsulate modern rule-based and duty-based ethical concepts appropriate to the engineering enterprise. They aimed to guide the actions of engineers toward doing the morally right thing in their employment of engineering methodologies. Within this type of ethics, the rightness of an action takes priority over the goodness of the agent. In contrast, in this chapter, I will emphasize virtue ethics, which links directly to character or *ethos* of the agent. I advocate restoring virtue ethics not to replace modern systems of process ethics, whether consequentialist or deonotological, but to complement these ethical theories.

ETHICS AND MORALITY

Ethical theory informs moral practice. Moral practice grounds ethical theory. One can, for instance, speak of moral theories abstracted from grounded concrete conduct. This, I believe, is one way virtue ethics is developed. One can speak of conceptual ethical theories deduced from abstract ethical principles. Ethical theories, particularly theories of virtue ethics, as practical theories, might be seen as mid-range phenomena, lying between the concrete and the abstract.

German idealist philosopher Georg Wilhelm Friedrich Hegel (1770–1831) proposed a kind of practical reasoning grounded in what he calls *Sittlichkeit*, which can be translated as "moral substance" or "moral life," the practical realm of moral being and conduct. It is the world of our everyday involvements in which we encounter reality directly, bearing always already some sense of how to be in accord with "the ought" or "the should." Robert Piercey views a *Sittlichkeit* as a concrete historical community that has a shared way of life. "It has its own conception of duties, virtues, and the good, conceptions which are taken for granted in this community but which need not be accepted in others. From this standpoint, to be a good agent is to be a good member of one's *Sittlichkeit*—to know one's station and one's duties."[4] The moral life of one's *Sittlichkeit* is contextual and communal and corresponds to Dewey's notion of customary morality. *Sittlichkeit* also resonates with the notion of *Gemeinschaft* in Ferdinand Tönnies's *Gemeinschaft/Gesellschaft* distinction referring to two fundamental types of human association. *Gemeinschaft* is restricted and exclusive but heart-felt and familial, whereas *Gesellschaft* is objective and reflective but more inclusive and open.

Now, even though moral reflection within one's *Sittlichkeit* appears, at best, as a vague pursuit, contemporary philosopher Paul Ricoeur believes we can flesh out our indefinite intuitions by following Aristotle and explicating moral reflection that aims at a "good life," with others and for others and within just institutions.[5] More precisely, we must first of all inquire about the good life and take steps toward realizing it. Second, we must do so in the context of being social animals, even though we might prefer to keep our own company. Third, we must proceed with fairness toward others we encounter within the historically conditioned social institutions we find ourselves thrown into. These are the tasks, according to Ricoeur, to which we humans are assigned in order that we might flourish in the fullness of our moral experience. These are the first steps in development of a theory of virtue ethics, and they correspond to a movement away from one's *Sittlichkeit*.

But if we move away from *Sittlichkeit*, toward what do we move? Hegel distinguishes *Moralität* from *Sittlichkeit*. *Moralität* is the realm of practical ethical judgments, based largely in the philosophy of Kant, within which a general notion of duty is paramount. The notion of *Moralität* insists that the duties we have in common have nothing to do with the particulars of our ethical communities, or even with anything in the phenomenal realm, because pure reason gives a moral law to itself, and this law obliges unconditionally. It is always and everywhere valid.[6] Hegel had in mind specifically Kant's deontological moral theory of duty, which

4. Robert Piercey, "Not Choosing Between Morality and Ethics," *The Philosophical Forum* 32, no. 1 (2001): 54.
5. Ibid., 59.
6. Ibid., 54.

I looked at earlier in the text. *Moralität* is essentially the realm of ethical principles and the practical judgments stemming from them. It resonates with *Gesellschaft* in the *Gemeinshaft/Gesellschaft* distinction.

Consequentialism, as discussed in Chapter 2, is another form of ethical theory that would be at home in the realm of *Moralität*. The most familiar form of consequentialism is utilitarianism, which has wide circulation in today's world. Ethical theory comes in many guises. Utilitarian ethics assumes that one should always seek the greatest good for the greatest number of people. It would, for example, not require that a promise be kept if the greater good can be achieved by breaking it.

According to John Rist, Hegel was wary of an overreliance on these rule-based systems and this realm of principle, this *Moralität:* "Any moral system based on references to rules and principles alone (and which thus discounts the emotions and intuitions, including empathy, as well as the proper *satisfaction* in doing good both immediately and habitually) is *unlivable* and necessarily leads either to hypocrisy or to the abandonment of morality itself."[7] In addition, as conceptually interesting and coherent as these rule-based systems might be, they have lately come into question, not because they are not useful, but rather because they are not used. In fact, as I mentioned in Chapter 2, one study concluded that most engineers do not explicitly follow a code or system of ethics when responding to moral situations; they respond instead in an intuitive, spontaneous way in accord with their character or *ethos* from within their particular *Sittlichkeit*.[8]

Is Hegel's *Sittlichkeit/Moralität* distinction the same thing as the ethics/morality distinction? If we envision ethics as a theoretical concern with principles developed from moral practice, then we can say that ethics is more at home in the Hegelian realm of *Moralität*. And if we envision morality as actual human conduct in the world of our everyday engagements, then we can say that morality is more at home in the Hegelian realm of *Sittlichkeit*. I think it is useful to view the *Sittlichkeit/Moralität* distinction along a continuum. At one end, there are highly theoretical and abstract ruminations about "the ought," and at the other, there are totally concrete and experiential practices. In between, there is experiencing, being, doing, and multiple forms of thinking: abstract thought,

7. John Rist, *Real Ethics* (Cambridge: Cambridge University Press, 2002), 120.

8. Roy V. Hughson and Philip M. Kohn, "Ethics," *Chemical Engineering*, September 22, 1980, 132.

more practical thought, common sense, intuition, and so forth. Matters of ethical and moral import participate in a dialectical or hermeneutical oscillation of the elements of the *Sittlichkeit/Moralität* continuum. Theories of virtue ethics, developed within the continuum, are practically oriented. Concrete agents of virtue ethics, premodern engineers in our case, practiced their particular virtues within a familiar and everyday world, their *Sittlichkeit.*

THE VIRTUOUS AGENT

When agents, engineers, act intuitively in a virtuous manner, we say they are acting in accord with customary morality. But what we are after in this chapter is Dewey's notion of reflective morality. How can the ancient ethics of virtue be seen as an instance of reflective morality? Its affinity for the affective life puts a theory of virtue ethics toward the middle of the *Sittlichkeit/Moralität* spectrum. Julia Annas maintains that ancient ethical theories are centered on the life of the agent as a whole. Concern with character, choice, practical reasoning, and the role of the emotions is at the heart of these theories.[9] Still, theories are only theories, and as Annas cautions us, "All theories, whether deontologieal, consequentialist, or virtue-based, direct us toward the principles we need to apply to produce right answers, but this does not mean that they give us the answers; *we* apply them to give the answers."[10] That application and the interpretation of the results that follow, along with the development of the theory itself, are precisely the tasks of the virtuous agent acting in accord with reflective morality.

Theories of virtue ethics do not typically follow the model of scientific theory, which involves abstraction, reduction, and formulation of principles from which practice proceeds. Theories of virtue ethics do not tend to be hierarchical and complete. They evolve more as a set of notions that includes, primarily and primordially, the agent's final end, the *telos.* They involve, as Ricoeur claims, aiming at some notion of the "good life," engaging with others and being for others, and working out details in the context of a just institution, which for us is engineering. Although there is general agreement that our general *telos* is happiness (*eudaimonia*), the

9. Julia Annas, *The Morality of Happiness* (Oxford: Oxford University Press, 1993), 4.
10. Ibid., 6.

nature of that happiness is a wide open topic, which requires an enveloping conversation.

The framework of particular virtues, of concern to a theory of virtue ethics, is generally nonhierarchical. Scientific theories postulate general principles, which attempt to subsume particular cases in a clear and distinct fashion. Theories of virtue ethics embrace the character of the person/agent/engineer in the wholeness of their historical being, and unlike scientific theories, they function more like a web of relationships thrown over a sector of reality that structures and shapes that reality.

So how do virtuous agents, like engineers, for example, plying their trade within the institution of engineering, put into play a virtue ethics? The principles of a virtue ethics would be the virtues themselves and their advocacy. How are the virtues to play a role in the formation of the character or *ethos* of the traditional engineer? Joel Kupperman maintains that, in general, character is a web of involvements which includes the presence or absence of (1) dispositions to recognize certain situations as ethically problematic, (2) dispositions to treat certain factors as having special weight in ethical decisions, (3) concerns for certain things thought to matter, and (4) commitments that provide a connecting thread among different moments of the agent's life.[11] Commitments, concerns, and dispositions—as well as the volition needed to put these ways of being into play—are all involved in the formation of *ethos* or character fundamental to the development of a virtue ethics.

A framework for a theory of virtue ethics might consist of an initial *ethos,* a desired final *ethos,* and virtues to be practiced as means to get from the one to the other. Greek ethics, the quintessential type of virtue ethics, had a theoretical structure consisting of three moments. Initially, William J. Prior maintains, "There is an ideal form of human existence, which individuals strive to attain and which it is possible for them to achieve. Second, the virtues are necessary to the attainment of this ideal. Third, wisdom or knowledge is at least one of the virtues needed for its attainment."[12] The wisdom I have in mind is the *practical* wisdom, or *phronesis,* recommended by Aristotle. There is another design for a theory of virtue ethics from Alasdair MacIntyre. His structure consists of the

11. Joel Kupperman, "Character and Ethical Theory," *Midwest Studies in Philosophy* 13 (1988): 116.

12. William J. Prior, *Virtue and Knowledge: An Introduction to Ancient Greek Ethics* (London: Routledge, 1991), 4.

three questions: who one is, who one ought to become, and what form of action will bring that person from the present to the future.[13]

Now, although it is common to say that a virtuous person is a person of character, I think it is important to keep clear the difference between character and virtue. Character is who we are in terms of concerns, commitments, dispositions, and volition. Virtues quite simply are those dimensions of a person's character that have moral worth, over which a person has some freedom of choice. Persons can, in other words, be held responsible for the exercise or failure to exercise their virtues.

I am assuming that the virtuous agent of the premodern era was primarily moral, rather than ethical, simply because ethical systems as conceptual frameworks, developed within Hegelian realms of *Moralität,* only came into being with the rise of modernity. In recent years, within the postmodern turn, we have witnessed a decline in the authority of rule-based or law-based ethical systems. Does this mean that in the future ethics will disappear? As Zygmunt Bauman explains:

> It may well be that the power-assisted ethical law, far from being the solid frame which protected the wobbly flesh of moral standards from falling apart, was a stiff cage that prevented those standards from stretching to their true size and passing the ultimate test of both ethics and morality—that of guiding and sustaining inter-human togetherness. It may well be that once that frame has fallen apart, the contents it was meant to embrace and contain will not dissipate, but on the contrary gain in solidarity, having now nothing to rely on but their own inner strength. It may well be that with attention and authority no more diverted to the concerns with ethical legislation, men and women will be free—and obliged—to face point-blank the reality of their own moral autonomy—and that means also of their own non-get-riddable, inalienable moral responsibility. It may happen (just may) that in the same way as modernity went down in history as *the age of ethics,* the coming post-modern era will come to be recorded as *the age of morality.*[14]

13. James F. Keenan, S.J., "Virtue Ethics: Making a Case as It Comes of Age," *Thought,* vol. 67, no. 265, June 1992, p. 120.

14. Zygmunt Bauman, *Life in Fragments* (Oxford: Blackwell Publishers, 1995), 36–37.

PRACTICE

Virtue must be practiced. A good character does not emerge from isolated good acts or from a natural inclination to lead a good life. A good character emerges only through more or less explicit adherence to a virtue ethic. Engineering, of course, consists of acts, and moral judgments concerning acts fall typically within the domain of rule-based ethics, whereas virtue ethics focuses on the character of the agent involved in the practice. Within the practice of engineering, then, virtue ethics and rule-based ethics tend to converge. Hence, codes of ethics in which rules, standards, values, and ideals of the engineering profession are most explicitly formulated should reflect the virtues that indicate the morally good engineer. An engineer, for instance, who follows a code of ethics should become virtuous, while a virtuous engineer would no doubt follow the code of ethics of his or her discipline.

Julia Annas maintains that virtue is not a neutral kind of excellence or an excellence in achieving some non-morally specified way of being. Virtue is "a complex disposition to do the morally right thing for the right reason in a consistent and reliable way, in which one's emotions and feelings have so developed as to go along with one's decisions."[15] Dispositions, however, need to be enacted in order to be manifest. So, we can take virtue in a more dynamic sense as a practice in search of a moral excellence, which, of course, presupposes a disposition to engage in the practice in the first place. Alasdair MacIntyre defines virtue as "an acquired human quality the possession and exercise of which tends to enable us to achieve those goods which are internal to practices and the lack of which effectively prevents us from achieving any such goods."[16] These practices involve interpersonal relationships and social cooperation.

Crucial to MacIntyre's analysis is the distinction between internal and external goods. Money, for example, is an external good that might come from the practice of baseball. But an external good is a good that can be achieved by engaging in other activities as well, like making good real estate investments. On the other hand, there are goods internal to the playing of baseball that cannot be explained, experienced, or understood apart from the specific context of the practice. For example, becoming an

15. See Annas, *Morality of Happiness*, 441.

16. Alasdair MacIntyre, *After Virtue*, 2nd edition (Notre Dame: University of Notre Dame Press, 1984), 191.

excellent line-drive hitter is an internal good in baseball.[17] An example from engineering of an internal good: becoming an expert or even a virtuoso in the design and implementation of digital networks for a wide variety of engineering problems.

Having that proficiency is a good in itself in the sense that it embellishes the practice of engineering, advancing its benefits to society at large. In the process of acquiring that skill the engineer engages with others in a variety of cooperative ventures. It is precisely within that engagement that virtues are exercised. Of course, an engineer can use this internal good to achieve an external good, like fame, by presenting papers at conferences or writing textbooks.

The *ethos* of engineering is the *ethos* of engineers practicing as engineers. And that generally means practicing collectively. This is not to belittle the achievements of individual engineers. Their acquisition of skills is an essential element in both traditional and modern engineering practice. The point is that their technical skills as well as moral virtues are never truly possessed unless they are exercised and that exercise is invariably a collective enterprise. The collective practice of engineering must be distinguished from the institution of engineering. The collective practice aims at internal goods. The institution aims at external goods. Institutions form the necessary context for the practice. They are structured in terms of power and prestige. They distribute money, power, and status as rewards.[18]

The pursuit of external goods does not require the exercise of any special moral virtues, but depends on ambition and self-interest, natural human drives that institutions organize and channel. The problem is that even though the pursuit of external goods through its institutions is a necessary part of engineering practice, the pursuit of internal goods and the engagement of the virtues in engineering practice might bar us from being rich or famous or powerful. Thus, although we may hope that we can not only achieve the standards of excellence and the internal goods of certain practices by possessing the virtues and also become rich, famous, and powerful, the virtues are often a potential stumbling block to this comfortable ambition.[19]

17. Randolph Feezell, "Sport, Character, and Virtue," *Philosophy Today* 33, no. 3/4 (Fall 1989): 205.

18. See MacIntyre, *After Virtue*, 194.

19. See ibid., 196.

The possible conflict between internal and external goods is a serious concern, but the possible conflict between different types of internal goods is even more serious. The question of goods internal to engineering is complicated by the fact that the engineer is engaged in a practice that has various dimensions, which often exhibit value conflicts. Like many a modern engineer, physicist J. Robert Oppenheimer was attracted to a problem that to him was technically sweet but morally repugnant: nuclear weapons work. Oppenheimer was enthralled by nuclear science but was also deeply troubled about atomic weapons. He sought to delay the development of the hydrogen bomb while avenues for arms control could be explored.[20]

The question seems to be, How can engineers, while exercising technical virtuosity, design and create devices, networks, organisms, structures, and systems that influence society in ways that enhance the common or social good? The social good is an internal good of engineering practice that should take precedence over other forms of internal goods.

Morality stems from our connections with each other in shared worlds of our everyday commitments and concerns. Those connections between human and nonhuman beings take on myriad shapes, some of which are of particular concern to the engineer. The three kinds of connection I look at here are characterized by *being-toward* the other, *being-with* the other, and *being-for* the other.

Being-toward suggests a kind of objectivity. More completely, being-toward means finding myself being-along-side those other humans and nonhuman beings toward whom or toward which I have already adopted an attitude. I feel compelled to treat objectively and fairly the other toward whom or toward which I relate as an engineer. I may in fact not do so, but then I am not being virtuous. Being-with encompasses being-toward because it presumes being-toward but includes an additional realization that "we are all in this together." If so, being-with necessitates at least a sense of honesty. I feel compelled to treat in a spirit of openness and truthfulness the other with whom I relate as an engineer. Again, I may not do so, and may fail to be a virtuous engineer by being dishonest. Being-for the other encompasses both being-toward and being-with the other. In being-for the other, according to Zygmunt Bauman, we follow our most basic moral impulse. "Following the moral impulse means assuming *responsibility* for the Other, which in turn leads to the *engagement* in the

20. See Pacey, *Culture of Technology,* 125.

fate of the Other and commitment to her/his welfare."[21] The practice of care, objectivity, and honesty ensures the moral integrity of engineers engaged in the project of engineering and authenticates the place of that project within the human family.

In this discourse, I am talking about the practice of a system of virtues as seen from a contemporary perspective. The premodern engineer may certainly have practiced a different virtue set, perhaps involving virtues like temperance, loyalty, obedience, justice, and so on. It is important to keep in mind that the ancient form of virtue ethics was a rather different affair, not only in content but also in structure, than what we moderns are apt to embrace. This is because in the ancient world ethical theory was not viewed as a problem-solving mechanism. Ethical theory emerged from the need to reflect on ultimate goals in life and on how to live a happy life. A theory that led to solutions of ethical conflicts was secondary.[22] In the modern world, especially in the case of the practice of engineering, problem-solving is paramount, and we would no doubt like to have a virtue ethics that incorporates that fact. My intention is to lay out a virtue ethics that follows an ancient virtue ethics structure, providing a place from which an engineer can evolve into the fullness of his or her being, but at the same time involving an assessment mechanism—to be sketched out at the end of the chapter—that leans in the direction of the practices of a more modern agenda.

OBJECTIVITY

Engineers should practice the virtues of care, honesty, and objectivity in order that they might inculcate an ideal character in the service of the traditional values of the engineering endeavor. If all engineers are of authentic character, then the culture of engineering will be attuned to the good, and the virtuous practice of engineering will be in the service of the common good and the betterment of society.

The practice of virtues will no doubt be motivated by external goods of money, power, and fame; yet this motivation should be subordinated to the pursuit of internal goods. One reason I propose care, honesty, and

21. See Bauman, *Life in Fragments,* 100.
22. See Annas, *Morality of Happiness,* 443.

objectivity as fundamental virtues is that they encompass many of the proposals offered in the literature. For instance, Edmund Pincoff advises, for the modern world, and not necessarily just for engineers, that persistence, courage, and unflappability in face of setbacks are desirable instrumental virtues and tolerance and tact desirable diplomatic virtues.[23] These virtues, it seems, do speak to the needs of the engineer, and many of them fall under the broad heading of care and objectivity.

Another set of virtues, even more closely aligned with mine, comes from MacIntyre, who advocates the practice of the virtues of justice, courage, and honesty.[24] He sees these as being essential for governing the social relations involved in all our human practices. One more recent exposition of a system of virtues can be found in the book *The Moral Sense* by James Q. Wilson.[25] He suggests that we work on developing the innate sentiments, common to all humans, of sympathy, fairness, self-control, and duty. MacIntyre's notion of justice and Wilson's notion of fairness resonate with what I am calling objectivity.

Objectivity is the disposition of impartiality and all that that connotes. Following Lawrence Haworth, I see objectivity as roughly synonymous with a personal outlook or stance of impartiality or fairness, but with the added sense of being oriented outward onto the objects I encounter, locating meaningfulness in those objects themselves rather than in the way those objects impact me or the groups with which I identify.[26]

To be objective is not to be aloof, uninvolved, or uncommitted. It is to be disinterested rather than uninterested. To be objective is to take into account all relevant factors. To resolve an issue objectively or disinterestedly is to have as a trait of character a standing disposition to take into account only the facts, interests, and beliefs that are relevant to the issue at hand, setting aside one's biases and singular interests, deciding the matter on its own merits. We all have that disposition to some extent, but no one exhibits it consistently or totally.[27] My proposal is that engineers embrace this kind of circumspect objectivity as a contribution to a positive engineering character. Though most engineers are so inclined to

23. Edmund L. Pincoff, *Quandaries and Virtues* (Lawrence: University of Kansas Press, 1986), 6.

24. See MacIntyre, *After Virtue,* 191.

25. James Q. Wilson, *The Moral Sense* (New York: The Free Press, 1993).

26. Lawrence Haworth, *Decadence and Objectivity* (Toronto: University of Toronto Press, 1977), x.

27. Ibid., 100.

some degree, I want to designate this sense of objectivity as an explicit ingredient in the proposed virtue ethic. This kind of objectivity, practiced as a virtue, seems to provide the conditions that make possible an open-minded consideration of rule-based ethical systems, such as codes of engineering ethics. This notion of objectivity refers to a self-transcendence—a transcendence of the self, but not a transcendence of the project—within which I move beyond mere self-interest. I become more than myself by affirming my individuality but at the same time affirming my connection to the other beings in my world. The objective engineer practices in a socially responsible way. Such practice, says Haworth, "directs the person outwards, toward other persons and objects to which value is attributed, and the self affirmed through the activity is one defined by its relationship with those persons and objects"[28]

Here is an example. A positive engineering solution to the problem of land erosion calls for a practice of the virtue of objectivity toward grazing land. Many ranchers graze cattle in small fenced-in areas, because cattle are easier to control this way. The problem is that the cattle in such circumstances tend to consume all the green edibles in that area, which leaves nothing to hold the topsoil together, causing erosion of the soil. Agricultural engineers, working perhaps as a consulting team hired by the ranchers, might suggest to the ranchers that the solution to their erosion problem is to graze the cattle in larger areas. Cattle will then only eat the greenest grass and leave much foliage that will hold the topsoil together. Of course, this solution is expensive. Can ranchers afford it? Can they afford not to? The view of land as a resource that has value only to the rancher who owns it must be expanded "objectively" to be able to see the land as having value in its own right and to persons and animals other than the landowner. This is the "big picture" view that is fair and impartial. Such a view might decrease short-term profits to the owner and thus be seen to be in opposition to the pursuit of their external goods. Most ranchers are business people who aim to make money. The consulting engineers in my example are also in business to make money, but for them to act as true professionals seriously engaged in objective engineering practice, they must look at the erosion problem with the widest possible perspective. That invariably implies that they recommend

28. Ibid., 91–92.

environmental responsibility, the long-term approach, over the short-term profit motive. Trade-offs will be necessary. The conflict between short-term and long-term goals will be an inexorable presence in much professional practice. Nevertheless, the objective engineer, in the face of a variety of allegiances—to client, company, profession, and the social world—must practice in an environmentally and socially responsible manner, which may seem to be at odds with the pursuit of external goods, at least over the short term.

The element of self-transcendence as part of objective engineering practice also enhances the professionalism of engineering. The dedication of engineers to their profession implies that they are directed not only toward satisfaction of personal wants, but also toward expressing the values of the socially defined practice of the engineering profession. An engineer will ideally value and be directed toward the practice that holds the highest standards. These standards are passed on through the various ways that engineers learn to engineer within the social tradition of engineering practice. For example, even though formal apprenticeship programs do not exist as much today within engineering as within medicine, newly hired young engineers are usually called on to assist senior engineers in various projects. Also, codes of engineering ethics, reflecting and summarizing the social traditions of engineering practice, articulate what the tradition views as the responsibilities practitioners have to one another as participants in their practice. Treating each other fairly as codes of ethics mandate, for instance, implies the practice of an underlying virtue of objectivity in the sense discussed here.

Virtues are qualities necessary to achieve the goods internal to practices. Practicing the virtue of objectivity leads to a self-transcendence that opens up the context of engineering practice. Within that expanded view, we see that internal goods can only be achieved by subordinating ourselves within engineering practice to our relationship to other engineers. As practicing engineers we are always already in relationships to other engineers with whom we share the kind of purposes and standards that inform the practice of engineering.[29] Virtues are not only qualities necessary to achieve the goods internal to practices, but are also the qualities contributing to the good of one's whole life.

29. See MacIntyre, *After Virtue,* 91.

HONESTY

Nietzsche considered honesty the *only* virtue. In his *Autobiography,* Ben Franklin confesses that when he was young, he and his friends stole stones from a house about to be built. They used the stones to build themselves a little wharf for their fishing-hole. "The next morning the workmen were surprised at missing the stones, which were found in our wharf. Inquiry was made after the removers; we were discovered and complained of; several of us were corrected by our fathers; and, though I pleaded the usefulness of the work, *mine convinced me that nothing was useful which was not honest.*"[30] True utility, which is at the heart of the engineering endeavor, in fact, calls for a sense of honesty on the part of all involved. Even though Franklin suggested hard work and frugality as primary *individual* virtues required of us all, he declared honesty to be his primary *social* virtue. "Whereas effective people tend to be distinguished by their hard work and frugality, effective societies tend to be distinguished by their social commitment to honesty."[31] The society we have in mind is the society of engineers, working together, being-with each other, connected in a cooperative and truthful manner, not just to other engineers but also to people in the social world who motivate their endeavors and are impacted by their products.

Ephesians 4:25 commands, "Wherefore putting away lying, speak every man truth with his neighbour: for we are members one of another." In being-with we are certainly connected, each to each. However, sometimes the truth can be neither kind nor useful. Knowing when to speak and when to keep silent requires the integrative virtue of *phronesis.* As MacIntyre tells us: "Lutheran pietists brought up their children to believe that one ought to tell the truth to everybody at all times, whatever the circumstances or consequences, and Kant was one of their children. Traditional Bantu parents brought up their children not to tell the truth to unknown strangers, since they believed that this could render the family vulnerable to witchcraft. In our culture many of us have been brought up not to tell the truth to elderly great-aunts who invite us to admire their new hats."[32]

30. *Benjamin Franklin's The Art of Virtue,* 3rd edition, ed. George L. Rogers (Eden Prairie, Minn.: Acom Publishing, 1996), 137.

31. Blaine McCormick, *Ben Franklin's 12 Rules of Management* (Irvine, Calif.: Entrepreneur Press, 2000), 111.

32. See MacIntyre, *After Virtue,* 192–93.

For virtue ethics lying is of course wrong, but in a manner that is not as direct or as strict as it is for the deontological ethics of Kant, which declares any lying absolutely incompatible with moral rectitude. In virtue ethics actions are judged right or wrong according to one's overall character. Virtues are positive qualities of people—culturally ingrained on the one hand, but freely chosen on the other—which orient us toward certain types of behavior and not others. Objectivity and honesty, for instance, are virtues engineers might practice to help fulfill their human potential. From the perspective of virtue ethics, to practice the virtues is to be ethical. With virtue ethics it is difficult to assess, in an unambiguous manner, the morality of distinct acts. Nevertheless, virtue ethicists typically consider lying wrong since it is in opposition to another virtue, namely honesty. But there is ongoing discussion about whether a lie might be acceptable when told in the service of another virtue. There is the classical example of a conflict between honesty and compassion: the son's lie to his drunken father is in the service of compassion for the mother and her physical safety. Conflict between virtues is resolved by appealing to a notion of the unity of the virtues, which is really just another version of practical wisdom or *phronesis*. In accord with this unity, the ideal virtuous person achieves all the virtues whenever he or she achieves any one of them. What this might entail is to emulate an ethical exemplar, a person whose ethicality comes across as balanced, centered, focused, and neutral. In looking at lying from this broader perspective, virtue ethics would find it to be immoral only if it indicates a divergent movement away from, rather than a convergent movement toward, the process of coming into the fullness of our moral being within the context of being-with others.

To be with the other in a way that leads to a flourishing existence requires at the very least that we are civil and honest with one another. This is certainly true of engineers. As engineers in contemporary times, whether premodern or modern, working typically in teams, we strive to bring useful things into being, to realize the promise of technology. The need for trust and therefore honesty is amplified in such settings. Recall my earlier mention of the distinction between *Gesellschaft* and *Gemeinschaft* outlined by the German sociologist Ferdinand Tönnies (1855–1936).[33] *Gesellschaft* is characteristic of the being-with form of relationship,

33. Ferdinand Tönnies, *Community and Society: Gemeinschaft und Gesellschaft*, trans. Charles P. Loomis (East Lansing: Michigan State University Press, 1957).

Gemeinschaft of the being-for form. *Gesellschaft,* ordinarily translated as "society," is largely functional. The members of a *Gesellschaft,* such as a typical engineering team, are being-with each other. They must practice objectivity and honesty in their relations, but their bond is their common purpose, the project. The members of a *Gemeinschaft* have the potential of being-for each other. A long-standing premodern engineering community or a group engaged in focal engineering would find their bond in each other.

To practice the virtue of honesty in its purest form means to tell the truth, to avoid the lie. Honesty is a virtue we would no doubt like everyone to practice. It greases the wheels of social interaction. In the practice of science and engineering, lying can cause severe problems. As Edmund G. Seebauer and Robert L. Barry maintain, misrepresenting data or procedures derails the effort to discover the truth about reality, and inflating one's contribution to a project at the expense of others corrodes the team effort needed for long-term success.[34] In their seminal work on engineering ethics, Harris, Pritchard, and Rabins point out seven different ways that engineers can misuse the truth. These include intentional falsehoods or lies, deliberate deception, withholding information, failing to adequately promote the dissemination of information, failure to seek out the truth, revealing confidential or proprietary information, and allowing one's judgment to be corrupted.[35]

A classic example of honesty as well as misusing the truth is the case of the 1986 *Challenger* disaster, in which seven astronauts lost their lives.[36] The problem of the faulty O-rings is widely known within the engineering community. Morton-Thiokol engineers Roger Boisjoly and Arnie Thompson expressed their concerns about the booster rockets under cold weather conditions. They were trying to be virtuous engineers. The cold would exacerbate the problems of joint rotation and O-ring seating. Predicted temperature at flight time was to be 26°F, but the lowest temperature ever experienced by the O-rings in any previous mission was 53°F in a 1985 flight. Even at this temperature, the boosters had experienced O-ring erosion. Since the engineers had no temperature data below 53°F, they

34. Edmund G. Seebauer and Robert L. Barry, *Fundamentals of Ethics for Scientists and Engineers* (Oxford: Oxford University Press, 2001), 145. This is the first engineering ethics text that stresses virtue ethics.

35. Harris, Pritchard, and Rabins, *Engineering Ethics*, 119–21.

36. The information on the Challenger disaster was drawn from the CD-ROM included with Harris, Pritchard, and Rabins, *Engineering Ethics*.

could not prove that it was unsafe to launch at lower temperatures. The engineers recommended that NASA delay the launch. A management decision was called for, and after much haggling, management decided that from their "engineering assessment" a launch was recommended. These managers were not being honest. Of the seven types of dishonesty, this is a case where they were allowing their judgment to be corrupted, in this case for political reasons.

CARE

I have established that a virtuous engineer must practice objectivity and honesty; yet these alone are not sufficient. He or she must also practice the virtue of caring. If I am being-with a group of other humans, say my colleagues or my friends or even a group of people on the street, and practicing honesty in an intersubjective manner, I might want to further my connections by asking, What else is there? The next step beyond mere being-with is being-for, as propounded by philosopher Emmanuel Levinas. His thinking initiates from the idea that ethics arises out of relations with other people, not from a universal law.[37] The source of meaning in our lives, that to which we are responsible, is the friend or stranger in whose eyes I can recognize myself.

Caring for the other requires doing something that true being-for demands. To the helpless I must provide help. To the starving I must provide food, and so on. The care exercised in the being-for relationship opens up the possibility of an authentic being-with and being-toward exhibited in the practice of objectivity and honesty. Care makes us one. Care in the widest sense links humans to their world. I care about "x," and "x" is thereby incorporated into my world, my context of involvements, which in turn grounds and conditions me. Care is connective. It is the spontaneous enactment of the relationship of being-for the other. Just as being-toward may elicit a deeper relationship of being-with, so too can being-with give rise to being-for. Here is Zygmunt Bauman echoing Levinas:

> To take a moral stance means to assume responsibility for the Other; to act on the assumption that the well-being of the Other is a precious thing calling for my effort to preserve and enhance

37. Jill Robins, ed., *Is It Righteous To Be? Interviews with Emmanuel Levinas* (Stanford: Stanford University Press, 2001), 114–15.

> it, that whatever I do or do not do affects it, that if I have not done it, it might not be done at all, and that even if others do or can do it this does not cancel my responsibility for doing it myself. . . . As the greatest ethical philosopher of our century, Emmanuel Levinas, puts it—morality means being-for (not merely being-aside or even being-with) the Other. And this being-for is unconditional (that is, if it is to be moral, not merely contractual)—it does not depend on what the Other is, or does, whether s/he deserves my care and whether s/he repays in kind.[38]

The care I have for the other flows along the threads of the web of relationships. That web of life, claims Carol Gilligan, depends on connection and is sustained by activities of care. It is based on a bond of attachment rather than a contract of agreement.[39] Like Levinas, Gilligan detects a totally different approach to ethical life than acting on universal principles, which was Kant's approach. The "different voice" Gilligan wants us to hear is the voice of care, doing instinctively whatever the situation demands.[40]

Nel Noddings elaborates on Gilligan's ethics of care, providing distinctions and definitions for a philosophical analysis of care. Noddings distinguishes the care-giver as the "one-caring" from the recipient as the "cared-for," insisting that both parties must be involved in this relationship if caring is to be complete. Noddings also contends that our obligation to actually be caring is limited by the possibility of reciprocity. We are not obliged to act as one-caring if there is no reciprocity, no possibility of completion in the other.[41] This implies that ethical caring is restricted to the human domain. As we move from the human level to the animal level to the plant level to the level of the inanimate object and the idea, we move "steadily away from the ethical toward the sensitive and aesthetic."[42] We can have a kind of nonethical or aesthetic caring, Noddings suggests, for beings that are not members of the human race.

38. See Bauman, *Life in Fragments*, 267–68.

39. Carol Gilligan, *In a Different Voice* (Cambridge, Mass.: Harvard University Press, 1982), 57.

40. Hubert L. Dreyfus and Stuart E. Dreyfus, "Toward a Phenomenology of Ethical Expertise," *Human Studies* 14 (1991): 246.

41. Nel Noddings, *Caring: A Feminine Approach to Ethics and Moral Education* (Berkeley and Los Angeles: University of California Press, 1984), 149.

42. Ibid., 161.

But this seems a bit contrived, a move by Noddings to save her notion of reciprocity. Levinas and Bauman, in fact, argue just the opposite, that the other's mere presence obliges us to be there for him/her/it. Deane Curtin, writing from the ecofeminist perspective about the treatment of nonhuman animals and the environment, generalizes and expands the idea of care advocated by Noddings by distinguishing between caring about and caring for.[43] I care *about* a general idea. I care *for* a specific person. And I can care about things as well as ideas. My involvement with a rock, an idea, a tree often involves an ethical dimension, an "ought." The ought of environmental ethics, for instance, insists that we listen to the voice of nature.

Though the needy other may oblige me in such a way that I respond without thinking of any kind of reciprocity, there is still a kind of reciprocity, or at least connectivity, at work among humans and between humans and the nonhuman world. I project my possibilities onto an "x" that I encounter. But to live in fullness of the experience of "x," I ought to embrace the possibilities that "x" projects onto me, to let "x" speak in a sense. This field of mutually projected possibilities is the world of care, and this world includes humans I care for from whom I can hope for—but not require—some kind of reciprocity, as well as things I care about, which from time to time grace me by the presence of their being.

To be sure, we humans are caring creatures in the sense of being connected to our world, typically interacting with a variety of other humans, pursuing goals, enacting roles, and employing an interconnected array of tools, devices, structures, and systems. The question becomes, How might these ideas of the broad sense of care speak more directly to a possible virtue ethic for engineering? Milton Mayeroff's slim volume *On Caring* lacks the philosophical depth and political urgency of the works of Gilligan, Noddings, and Baumann; nevertheless, his view of the caring relationship is very specific and systematic and seems especially relevant to the world of engineering. Following Mayeroff, I will view care as the practice of helping the other to grow, where "other" can refer to people, places, projects, ideas, or things. I experience what I care for (a person, an ideal, an idea) as an extension of myself and at the same time as something separate from me that I respect in its own right.[44] The idea of something separate from me that I respect in its own right recalls the virtue of objectivity

43. Deane Curtin, "Toward an Ecological Ethics of Care," *Hypatia* 6, no. 1 (1991): 67.
44. Milton Mayeroff, *On Caring* (New York: Harper & Row, 1971), 5.

as treatment of the other with fairness. In caring for the other I do not insist on my own agenda, but I appreciate the needs of the other, and I follow the direction of the other's growth. Generally, Mayeroff maintains, I experience the other's development as bound up with my own sense of well-being. How I am and how it is are mutually interdependent because, in the context of care, I am always already in the world, in relationship with an array of others.

Mayeroff's notion of care is very broad and subsumptive. It is synonymous with mutual concern, benevolence, and the kind of love that was known as *caritas* in the medieval world. St. Thomas viewed *caritas* as the greatest of the virtues. St. Augustine maintained that all we needed to do was practice charity (love, care), and the good life would automatically fall into place: "dilige et quod vis fac" [love and do what you will]. Mayeroff's notion of care implies a host of other virtues: knowing (implicit as well as explicit knowing), patience, trust, humility, hope, courage, and what he calls alternating rhythms (moving back and forth between a narrower and wider perspective on a given problem). Mayeroff also mentions honesty as involved in his notion of care, although I treat it separately. Also, as I mentioned, the virtue of objectivity, considered in the comprehensive manner presented previously, resonates with many of the virtues that Mayeroff's notion of care implies.

Care, enlivened by and enlivening the creative spirit, is at the heart of the endeavor of being human, and that creative spirit is manifest to varying degrees in all aspects of life, including the practice of engineering. Though some might claim we need to sever our connections in order to do good engineering, it is only those connections that allow us, in the first place, to comprehend the needs that provide the impetus for the engineering project. A certain kind of disconnection, a de-worlding, was seen to be intrinsic to the modern engineering enterprise. But, as Andrew Feenberg maintains, such de-worlding will be complemented by a world-disclosing process that requires a caring connectivity.[45]

How do we in doing engineering invoke the virtue of care? I think the character of most engineers is always already attuned to care to a large degree. To increase the stature of care we need only invoke it more explicitly. As mentioned earlier, there are stories of some heroic engineers

45. Andrew Feenberg, "Modernity Theory and Technology Studies: Reflections on Bridging the Gap," in *Modernity and Technology*, ed. T. Misa, P. Brey, and A. Feenberg (Cambridge, Mass.: MIT Press, 2003), 73–104.

putting the resources and expertise of their small computer companies at the service of a community center in a poor part of town, or others going to third world countries and setting up water procurement systems that immensely improve the quality of life for obviously needy people.

PHRONESIS

The virtues, however, are characteristics of persons, of the character of persons, not algorithms for solving ethical problems. Problem-solving requires another human excellence, the perfection of the activity of deliberation, which is called the virtue of prudence, practical wisdom, or *phronesis*. It is the excellence of discerning what is truly good for us, as individuals, or for human beings in general. It is the excellence of a certain kind of thinking—what Aristotle called an intellectual virtue. However, *phronesis* plays an instrumental role in guiding action to its proper end. It involves both theory and practice. Aristotle differentiated *phronesis* from, among other things, theoretical wisdom (*sophia*) and scientific thinking (*episteme*). It is through *phronesis* that we discern and choose appropriate goals of ethical virtue. Thus, ethical virtue without *phronesis* remains directionless. But discernment of the good and perfection of deliberation are dependent on having a good character. Hence, without ethical virtue, one might be able to figure out an appropriate end, but one would not be able to choose the proper means. And without *phronesis*, one might be able to choose the appropriate means but not the right end. Excellence of character, then, and practical wisdom together form a unity.

The virtue ethic I propose, headlining care, honesty, and objectivity, should promote an ideal ethos for the engineering profession. Tying them all together, achieving the unity of these virtues, requires *phronesis*. How does one take on such an ethic? *Phronesis* seems to take time and the incorporation and integration of the experience that goes with it, but to get started, no complex equations or calculations are required. (As I will show in the next section, however, some *simple* equations might be helpful in initiating the ethical discussion.) The most direct approach is simply to begin to be and act in a more caring, honest, objective way. With practice and experience these virtues will become part of the fiber of the being of the engineer. Engineers, practicing objectivity, honesty, and care over time, develop the soundness of judgment we call *phronesis*. There are engineers who have a knack, an intuition, for seeing the technically right thing to do even in very complex technical situations. What I am calling for is more ineffable, more rare, more difficult to discern. I am calling for engineers who have a knack, an intuition, for seeing the morally right thing to do even in very complex *moral* situations.

Roger Boisjoly's experiences as a Morton-Thikol engineer involved in the Challenger space-shuttle flight would, I believe, qualify him as a *phronemos* (one who practices the virtue of *phronesis*). He looked at the O-ring quandary in a fair and objective manner. There were obvious problems in that performance data was simply unavailable at low temperatures. Boisjoly was honest in his assessment of the situation. He forthrightly stated his belief that the flight should be delayed. He showed that he cared about the safety of the flight and for all involved. He took his position in the face of management pressures to give the flight a green light. Though management ultimately prevailed and the disaster that resulted shook the nation, throughout it all, Boisjoly acted as a true *phronemos.* Though his authority was limited, he acted as best he could and made his decisions, based on many years of experience, with the aim of contributing to the overall common good. He can be characterized as a *phronemos,* not just because he practiced the moral virtues of care, objectivity, and honesty, but also because he directed those virtues toward a proper goal, the common good, the health and safety of all involved.

Practical wisdom has a broad focus. It is proactive as well as reactive. Practical wisdom in the practice of engineering in a caring, honest, and objective manner can benefit the person as engineer and distinguish the engineer as person, as well as provide a positive focus for the profession. In order to practice the specifics of a particular task in harmony with a positive virtue ethics, we must aim not merely at external goods like money, power, and fame, but also at internal goods like virtue or social consciousness, sought in a collaborative engagement. Seeking such internal goods in a caring, honest, and objective manner should help to ennoble engineering practice, as the practice exhibits a more salutary influence upon the society in which it is embedded.

BRINGING IT TOGETHER AGAIN

In Chapter 2, I discussed a quasi-mathematical approach to initiating the discussion about process ethics. Here I will present an analogous approach for virtue ethics. To bring all the pieces of the puzzle together, I will consider all three virtues of objectivity, honesty, and care in terms of an example, the same example, in fact, that I used in Chapter 2 involving the engineering of Radio Frequency IDentification (RFID) devices. Again, assume the technical aspect of the engineering has been impeccably done, meeting all the standards of efficiency and productivity. How might we assess the moral dimension of the engineers involved in the engineering of RFID devices? We could certainly gauge them against the virtues I have put forth in this chapter to constitute a virtue ethics of the engineer, namely, the practice of objectivity, honesty, and care. This assessment can be carried out at the company level, perhaps initially within the group. Ultimately, assessments of the engineers involved in the RFID group will need to be done within the conversation of the lifeworld, with all interested parties involved.

If I assign a value function (J_o, J_h, J_c) to each of the three virtues I am aiming at, I can write

$$\text{By definition } J_v = \beta_1 J_o + \beta_2 J_h + \beta_3 J_c$$
$$\text{with } \beta_1 + \beta_2 + \beta_3 = 1.0$$

where the β_i terms are weighting factors whose values are to be determined by consensus. Initially assume all three value functions are equally weighted, so all β_i terms will be set to 1/3. Assume all three value functions can range from −3 to +3.

Assume the group leader is assessing the ethics of the engineers in terms of the virtue ethics discussed in this chapter. She decides that the objectivity or fairness of the RFID team members is at issue. With regard to such issues, Seebauer and Barry distinguish between different ways credit gets distributed. "Rewards include intangibles like notoriety and influence, and tangibles like promotion and salary increases."[46] Assume everyone on the RFID team is poised to get a ten percent raise because the team made a big breakthrough, but not everyone on the team contributed equally. Some members feel they did most of the work, and a few who did very little want all the credit. Seebauer and Barry: "Clearly, fairness requires that all members of the group contribute to the effort in proportion to the time and talent available to them. An open, honest discussion at the outset of the effort of who can best contribute what helps

46. See Seebauer and Barry, *Fundamentals of Ethics*, 175.

to avoid misunderstanding."[47] The group leader vows to have more open and honest discussions with her team members in an ongoing fashion and to keep closer tabs on the them in future engagements, but in an attempt to do a fair assessment of the current situation, she set the value for the fairness of the group to each other (J_o) at −1.5.

The honesty, let us say, of certain group members was also brought into question. That sent a negative ripple throughout the group. In assessing the overall honesty of the group and their apparent lack of care for each other, the group leader assigned J_h and J_c values of −1.5 as well. In retrospect, the team leader was surprised that the team did so well on the technical side, because their virtue ethics profile was not so good. The total value function J_v is computed:

$$J_v = 1/3(-1.5) + 1/3(-1.5) + 1/3(-1.5) = -1.5$$

indicating that some work needs to be done.

Again, this number is not the "answer" to the ethical question about whether or not these engineers involved in RFID device design, development, and manufacturing are virtuous. But the number can provide a point of departure for further discussion. The team can get together and decide that even though they were able to do good technical work with a not very good ethical assessment, they could in fact do even better technical work, and it will be more enjoyable for all concerned, if all team members were to strive for excellence in fairness, honesty, and care in their everyday human interactions.

Ultimately and eventually, the virtue ethics assessment profile can help to orient the discussion about engineering ethics at the wider level, as it moves beyond the personal level (discussed in this chapter) and the professional level (discussed in Chapter 2) into the conversation of the lifeworld at the social level.

CONCLUSIONS

In this chapter, I have laid out a general view of virtue ethics as it could be applied to the modern engineer, or even the premodern engineer in contemporary times. Exercising these virtues requires being-toward, being-with, and being-for other people, places, and things. I propose objectivity

47. Ibid.

or fairness or personal justice as essential in guiding the orientation of being-toward, honesty as essential to being-with, and care as essential to being-for. In general, the practice of a virtue ethic strengthens, balances, focuses, and centers all aspects of human being-in-the-world.

The loosely structured theory of virtue ethics I am putting forth here lies somewhere midway along the *Sittlichkeit/Moralität* spectrum. According to that theory, the engineer needs to be of good character. That is to say, he or she should be a "good" engineer—both technically and morally. Education, training, apprenticeships, and experience lead to technical proficiency, but moral proficiency requires the engineer to practice the virtues of fairness, honesty, and care in the pursuit of goals selected or approved by practical wisdom, or *phronesis*.

SIX

CONTEXTUALIZATION

In Chapter 3, I looked at engineering as a colonizing project. The project of the modern engineering enterprise colonizes the human lifeworld by imposing upon it the values of efficiency and productivity. That discussion assumes engineering is a decontextualized project that comes up with products of various sorts that impact the lifeworld. The true contextualized nature of the engineering project was bracketed in order to focus on colonization. In this chapter, I will make my analysis more comprehensive by discussing the contextualized nature of the engineering project without the need to bracket the colonization effect of it. Context, which the modern engineering enterprise sidesteps whenever it can, is more fully engaged within the premodern engineering endeavor. My working premise is that everything at least initially occurs within a context in both modern and premodern engineering.

In Chaper 3, I looked at engineering in its *part* aspect. Engineering was seen to be that particular project, looked at in terms of structure and function, which impacted the human lifeworld by colonizing it. In this chapter I will look at engineering in its *whole* aspect, engineering as a contextualized endeavor that has colonizing tendencies. The separated parts of engineering integrated into the systemic engineering project impact the holistic lifeworld while that lifeworld simultaneously contextualizes that engineering project.

What is context? Simply the surrounding world within which what is at issue comes to be at issue. One place where context plays a major role is in languages. American anthropologist Edward T. Hall distinguished between high-context and low-context cultures. He maintained that many Asian nations (China and Japan, in particular) can be categorized as nations with high-context cultures, and many Western nations (Germany, Britain, and the United States, for example) possess low-context cultures.[1] The meaning of what is said in a language and the context in which it is expressed, according to Hall, are inextricably bound up with each other. The question is how much meaning is found in the code of the language itself versus the context in which the language is spoken.

In many low-context cultures, such as ours, a minimal amount of the message meaning is embedded in the context, and more meaning is placed in the language code, or message. For this reason, messages in such low-context cultures tend to be more specific and detailed than messages in high-context cultures, where meaning resides primarily in the context. Most of the information in high-context cultures is either in the physical context or internalized in the person, and a minimal amount of information is transmitted in the message itself. Thus, to grasp the full meaning of a message, the listener must be able to decipher contextual cues. In elaborating on the work of Hall, R. S. Zaharna argues that more is expected of the listeners in high-context cultures than in low-context cultures. "When talking about something that they have on their minds, a high-context individual will expect his interlocutor to know what's bothering him, so he doesn't have to be specific. In other words, in high-context exchanges, much of the 'burden of meaning' appears to fall on the listener. In low-context cultures, the burden appears to fall on the speaker to accurately and thoroughly convey the meaning in her spoken or written message."[2]

Concerning the project of engineering, we tend to view the modern engineering enterprise as a "low-context" phenomenon. Whenever a contextual problem arises within engineering, we usually thematize that problem and make it part of the engineering project itself. For instance, the environment becomes problematic, and we incorporate EPA guidelines into our design processes. My aim, however, is to expand the typical purview of the engineering project and look at it as a "high-context" phenomenon. That means that context itself is given a voice and is not immediately incorporated into the engineering project. In other words, what would context look like if it were not colonized in a totalizing kind of way by

1. Edward T. Hall, "Context and Meaning," in *Intercultural Communication: A Reader,* ed. L. Samovar and R. Porter (Belmont, Calif.: Wadsworth, 1982), 18.

2. R. S. Zaharna, "Rhetorical Ethnocentricism: Understanding the Rhetorical Landscape of Arab-American Relations" (paper presented to the *Speech Communication Association,* 1995); available at http://academic2.american.edu/~zaharna/rhetoric.htm/.

the engineering project? The meaning and the place of the engineering project, I submit, would come more fully to light. Contextualization discloses. Decontextualization, as I have mentioned, is intrinsic to the modern engineering enterprise, while the premodern engineering endeavor—the principle focus of this second part of the text—has been a more contextualized affair. The language, we might say, of premodern engineering is high-context, and of modern engineering low-context.

Context is often paired off with content. Content is text. Content is embedded within context. *Con*-text is that which is *with* the text. The content of the engineering project was brought forth when I investigated the modern engineering process in Chapter 1. Recollecting some of those ideas, I might mention a typical sequence of steps one may see in the normal content of a modern engineering design process. For the premodern engineering endeavor, some of the following elements of content may apply, but normally in a much less clear and distinct fashion. The design team, whether modern or premodern, begins by gathering information and determining what the problem is. Then, if they are allowed to participate, they get together and do some soul-searching or brainstorming to determine what they ought to do. Then there might be a variety of steps, like modeling the system via mathematical equations, doing a sensitivity analysis of parameters within that model, investigating how compatible the various systems and subsystems are, doing a stability analysis of the system when subjected to various inputs, optimization of the design with respect to some preselected criterion, simulation and prediction of the performance of the system, and prototyping.[3] Following all these steps there is usually some kind of briefing or presentation of results. If the design is acceptable, a production stage is required, which often is designed along with the product itself. This process is called Concurrent Engineering, which is done to obtain optimized designs faster, at a lower cost, over the entire life cycle of the product.[4] Content, then, involving some or all of these steps in a typical design process, lies at the heart of the engineering project.

But the content, as text, is always embedded within context, the world

3. W. E. Wilson, *Concepts of Engineering System Design* (New York: McGraw-Hill Book Company, 1965), 5.

4. David Brazier and Mike Leonard, "Concurrent Engineering: Participating in Better Design," *Mechanical Engineering*, January 1990, 52.

of engineering practice. Again, context was crucial to premodern engineering and is less so to the decontextualized modern enterprise. But all engineering depends on context, even if it does not make an issue out of it. After all, it is only out of context that the need arises for the design in the first place, and back into context that the products resulting from the design are released. If content is figure, context is ground. Context grounds, conditions, limits, binds, modifies. And we always already find ourselves embedded within contexts, says Lorenzo C. Simpson, which are structured by interests, concerns, and fundamental commitments, and which give meaning to our practices.[5] Furthermore, continues Simpson, context "is the surrounding background of pre-understanding that frames our dealings with persons and things."[6] Using the dynamic metaphors of traffic and backdrop as substitutes for the more static figure and ground metaphors, we can say that these backdrops (contexts) against which a given content may traffic take many shapes and indicate a multitude of ways of being. The context of the project of engineering includes the spatial, temporal or historical, and conceptual backdrops against which the traffic of the engineering project comes into relief. Which dimensions of context, or which folds in the contextual backdrop, are crucial to the engineering project?

Context tends to be ignored, generally in modern engineering and understandably, if there is an urgency to get a product out the door quickly. Unfortunately, this mode of operation has become standard operating procedure for most modern engineering work. For the most part, modern engineering is business. When the heat is on, and nowadays that is pretty much all the time, it is hard to find time to worry about context. The realities of the marketplace, especially for the modern engineering enterprise, demand a decontextualization from the fullness of a world teeming with contingency. However, in order to gain understanding of the Big Picture of what we are about, we must pay attention to the meanings implicit in that context. For example, I know that this new cell phone my company is developing generates more power than most others on the market, and I also know that the danger of cell phone radiation is still an open issue. What do I do? Or what should I do? Economics, ethics, and environment are all contextual issues hiding in the folds of the backdrop of

5. Lorenzo C. Simpson, *Technology, Time, and the Conversations of Modernity* (New York: Routledge, 1995), 34.

6. Ibid.

my endeavors. Need they all be made explicit and factored into the structure of the content of the engineering process?

I am looking, then, for a general take on context and the way that context "contextualizes" the engineering project. What were the contexts of primary importance to the premodern engineer? The politics of the premodern era, particularly the ancient era, were no doubt repressive, and the engineer had to labor within a system of subjugation. Going along with the program would seem to have been a good idea, as it is still today for the most part. However, today's engineers tend to speak their minds quite openly, though less so in tight economic times.

The immediate context of the engineering project is the realm of technological systems, of which the engineering project is one among many, although through its products it is implicated in most of them. The realm of technological systems is embedded in the realm of systems in general, of which it is one among many. And the realm of systems in general consists in constructs arising out of the lifeworld: within the lifeworld there are systemic and nonsystemic ways to be. Martin Buber's famous nonsystemic I-Thou distinction is paired off against a systemic I-It.[7] The latter views the other as an object immersed in some kind of system, while the former views the other as a subject with its own intentionality. He was mainly concerned with the other as being a person but the other can be construed as being, e.g., nature. We take nature under the category of objective systemic reality and our relation becomes I-It. We take nature under the category of nonsystemic reality and our relation becomes a more heart-felt I-Thou.

The triple contextualization, including the contextualizing of systems, technological systems, and the engineering project itself by socially constituted decisions coming from the lifeworld, is indicative of one direction of the engineering/society relationship. A committee, say, of local citizens and technical experts gathers in the town hall (in the lifeworld) to discuss widening the highway that runs through town. They agree to form a subcommittee that will look at the matter in a systematic manner. Lifeworld contextualizes the realm of systems. After the subcommittee reports back in three weeks, the members of the lifeworld decide to go ahead with the widening effort. They ask local construction companies to submit bids on the project. To do this these construction companies have to study the

7. Martin Buber, *I and Thou*, trans. Walter Kaufmann (New York: Charles Scribner's Sons, 1970).

matter in more detail employing the apparatus of their technological systems. Lifeworld contextualizes the realm of technological systems. Finally, six months later, the bids are returned to the town hall gathering and, after much deliberation, the committee agrees on the best proposal. They hire one construction company, which employs a team of civil engineers and the engineering project gets under way. Lifeworld contextualizes the engineering project.

Another view of triple contextualization is that decisions made in the lifeworld contextualize the realm of systems. Then decisions made in the realm of systems contextualize technological systems. Finally, decisions made in the realm of technological systems contextualize the engineering project.

As indicated in Chapter 3, the triple colonization, that is to say, the colonizing of the lifeworld by the realm of systems, systems by technological systems, and technological systems by practices of the engineering project is indicative of the other direction of the engineering/society relationship: from the products of the engineering project to their social impact. But, as Feenberg, Dreyfus, and others indicate, not only does the lifeworld get colonized but it also gets *disclosed.* The heart of that disclosure is revealed in the decisions made within the conversation of the lifeworld.

LIFEWORLD

The lifeworld, as I have already mentioned, is the familiar world in which life is lived. It is so normal and ordinary that it is often overlooked. When I speak of the lifeworld, I will usually mean the nonsystemic lifeworld. Habermas generally views lifeworld being and systemic being as two separate issues. The lifeworld is the concrete, sensuous, participative, experiential context of ordinary everyday involvements. Herein people talk *to* each other, not just *about* each other. A healthy lifeworld is characterized by a human communication that exhibits mutuality and interdependence, as well as possibilities of genuine dialogue.[8] In the lifeworld, we generally try to comprehend each other and come to some kind of agreement. The two terms, comprehension and agreement, coalesce into something like communion.[9] The lifeworld is a place to develop one's identity amid a sense of belonging. The lifeworld is prescientific in that we know it by

8. Tod S. Sloan, "The Colonization of the Lifeworld and the Destruction of Meaning," *Radical Psychology* 1, no. 2 (Fall 1999): 4.

direct acquaintance, without needing theoretical grounding. This does not mean that in the lifeworld we are merely feeling things without any thinking going on. We do understand things in the nonsystemic lifeworld, but our understanding is what might be called primal or primordial understanding. Within the lifeworld we engage in spheres of action intersubjectively in seeking our mutual understandings. Once we begin to abstract this or that out of its larger frame of reference, we enter the realm of systems.

As Heidegger might put it: as we shift from the nonsystemic to the systemic ways of being, we shift from a *ready-to-hand* dealing with things in an instrumental totality to a *present-at-hand* dealing with things in a systemic totality. Within the instrumental totality of the lifeworld we relate to other humans immediately in terms of a *social integration;* within the realm of systems we relate in a mediated way via a *system integration.* In both the systemic and nonsystemic lifeworld we are—as I mentioned in the previous chapter's discussion of virtue ethics—being-toward, and being-with, and perhaps even being-for the other, where these ways of being are buttressed by the practice of a set of virtues: faith, hope, and charity perhaps, or maybe fairness, honesty, and care. Being in the lifeworld, either systemically or nonsystemically, with a disposition toward this or that set of virtues: that is how we find ourselves. But there are differences in the systemic and nonsystemic ways of being. The nonsystemic way of being in the lifeworld is a more free-flowing and less restrained way to be, although it is more vulnerable, because it is constantly threatened by systematization or colonization. Another way to say this is that we *belong together* with others in the lifeworld. However, in the nonsystemic realm of the lifeworld, the emphasis is on the *belonging,* and in the systemic realm, the emphasis is on the *together.*[10]

My notion of lifeworld, which should prove useful for investigation of the contextualized nature of the engineering project, is an amalgamation of ideas drawn from Heidegger, Husserl, and Habermas. Heidegger spoke of world, rather than lifeworld. For him the word "world" could be used in several ways. It could be used, he argued, to signify (1) the totality of

9. Klaus Hartmann, "Human Agency Between Life-World and System: Habermas's Latest Version of Critical Theory," *Journal of the British Society for Phenomenology* 16, no. 2 (May 1985): 149.

10. Heidegger uses the idea of Sameness in the sense of "belonging together" in "The Principle of Identity" (in *Identity and Difference,* trans. Joan Stambaugh [New York: Harper & Row, 1969]). "Belonging" indicates an intimacy between being and thinking, whereas "together" indicates a more customary scientific unification.

those entities of which we are conceptually aware; (2) the realm of possible objects for some sector of being, like the world of the engineer; (3) the place where humans live, like the public world which we share with a variety of others; or (4) the ontological and existential concept of worldhood.[11]

Clarifying Heidegger's fourth meaning of "world," Hubert L. Dreyfus says that worldhood "is the way of being common to our most general system of equipment and practices and to any of its subregions."[12] Heidegger said his normal concern was with the third sense of "world," which will be my general take on it too. We can view this sense of lifeworld simply as the context of our involvements, the place where we live and dwell and have our being. In another essay, Dreyfus maintains that our average everyday world is a public world, which we find to be "laid out in terms of roles, goals, and equipment for achieving these goals."[13] In a more recent work, Dreyfus, along with Charles Spinosa and Fernando Flores, elaborated on these three ideas, arguing that the world is a totality of interrelated pieces of equipment. It is an instrumental totality. Such a totality is used to carry out a specific task such as hammering a nail. Dreyfus, Spinosa, Flores: "These tasks are undertaken so as to achieve certain purposes, such as building a house. Finally, this activity enables those performing it to have identities, such as being a carpenter. These identities are the meaning or point of engaging in these activities."[14]

Premodern engineers as builders, for example, no doubt had at hand certain materials and tools to be used in road construction. Measuring apparatuses and digging tools were part of their set of equipment. They designed and constructed a road, employing a team of helpers; and if it was a good road, they became known as "road-builders," a title that bore some weight and lent meaning to their lives.

A modern engineer, as another example, steps into her laboratory, which is well stocked with the latest set of measuring devices: voltmeters, oscilloscopes, ammeters, function generators, and spectrum analyzers. Everything is properly connected to the operational amplifier circuit she has constructed. She applies the input signal and observes the output, a specific task in the service of designing a band-pass filter. She needs to

11. Heidegger, *Being and Time*, 97.

12. Hubert L. Dreyfus, *Being-in-the-World: A Commentary on Heidegger's Being and Time, Division I* (Cambridge, Mass.: MIT Press, 1991), 91.

13. Dreyfus, "Priority of *The* World," 121–22.

14. Charles Spinosa, Fernando Flores, and Hubert L. Dreyfus, *Disclosing New Worlds* (Cambridge, Mass.: MIT Press, 1997), 17.

achieve a certain bandwidth at a fixed resonant frequency. This filter circuit is to be used in a precision feedback control system for a disk drive. Her company is a leading provider of disk-drives for the computer industry. Her title of senior-disk-drive-engineer defines her work-world activities and gives meaning to this sector of her life.

Now, since Edmund Husserl was Heidegger's teacher, much of Heidegger's notion of world is a distillation of what he learned from Husserl. Husserl's idea of the lifeworld, as the world of lived experience, was most distinctly formulated in his *Crisis of European Sciences and Transcendental Phenomenology*,[15] published posthumously a decade after Heidegger's 1927 publication of his magnum opus *Being and Time*. In any event, Heidegger's world closely resembles Husserl's lifeworld, the world of immediate experience, which is always already there or "pregiven." Husserl distinguishes between a theoretical attitude toward life and a pretheoretical attitude, analogous to Heidegger's present-at-hand attitude and ready-to-hand attitude. Husserl's notion of the lifeworld or *Lebenswelt* refers to the everyday world of ordinary experience in which intuitions or summary understandings prevail. It is contrasted to the idealized, cognitive world of the sciences and mathematics. What makes getting a purchase on the lifeworld so difficult, by and large, is that this realm of immediate experience and intuition is generally attenuated even as it serves as the ground from which all sciences and theoretical development spring.

Habermas elaborates upon Husserl's and Heidegger's notions of lifeworld and world. His view is also based on the phenomenological lifeworld descriptions of Alfred Schutz.[16] For Habermas, as for Heidegger, actors in the nonsystemic lifeworld are not agents of a system, but are persons who find themselves thrown into a circumstantial world, being with others in a variety of ways, enacting a variety of ways of being. Habermas's lifeworld is a web of undoubted background beliefs and intuitions that act as a fund of meaning and a horizon for beliefs and practices that are more explicit.[17] Actors in the lifeworld, then, take on, or have already taken on, these beliefs and hold fast to them without, for the most part, bringing them to explicit awareness. They share experiences,

15. Edmund Husserl, *The Crisis of European Sciences and Transcendental Phenomenology*, trans. David Carr (Evanston, Ill.: Northwestern University Press, 1970).

16. See Hartmann, "Human Agency," 149.

17. David Kolb, "Heidegger and Habermas on Criticism and Totality," *Philosophy and Phenomenological Research* 52, no. 3 (September 1992): 687.

linguistic cues, and connotative understandings that allow participants in the process of communicative action to understand each other.[18] In the sphere of everyday communicative interactions, functions like child-rearing, education, and public debate transpire.[19] It is of course possible to view education, as in perhaps the conceptualized educational institution which employs a variety of educational methodologies, as being in the realm of systems rather than the nonsystemic lifeworld. Lorenzo Simpson points to another dimension of Habermas's structuring of the lifeworld, this time in terms of three components: culture, society, and personality or psychology. In its cultural dimension, the lifeworld is the realm wherein our practices take on meaning. In its social dimension, the lifeworld refers to normative orientations that facilitate group solidarity or social integration. In its psychological dimension, the concept of lifeworld refers to motivation structures that aid the acquisition of various speaking and acting competencies. In general, the lifeworld provides the backdrop against which, or the coordinate system within which, the traffic of communicative action unfolds.[20]

More specifically, what about engineers? They are certainly members of their lifeworld, as are we all. Engineers communicate in their lifeworlds with their friends and neighbors, coworkers and colleagues, as well as a plethora of other folks. This involves, as Klaus Hartmann maintains, the two actions of comprehension and agreement, which constitute what Habermas calls communicative action. Such "communing" unfolds within the conversation of the lifeworld. Because modern engineers aim to sidestep lifeworld and context in general, that conversation matters less to them than to premodern engineers.

Of course, part of this conversation involves the lack of comprehension and a variety of disagreements, as well as the struggles to understand each other and to arrive at consensus. Precisely out of the stew of this conversation come the ideas and decisions from which much of our social world is constructed, and that construction is often the result of mechanizations of the engineering project. Engineering, then, is contextualized by lifeworld decisions. Engineers have a major role to play in the conversation of the lifeworld, especially in light of the increasingly technological world emerging as a result of the work of engineers. Engineers have a better handle

18. Lambert, "Jürgen Habermas: Luddite Dragon or Defender of the Weak?" 4.
19. Feenberg, *Questioning Technology*, 167.
20. See Simpson, *Conversations of Modernity*, 81.

than most on the substance of these emergent technological phenomena. The decisions engineers make in the lifeworld as citizens of that world can be as crucial as the decisions they make in the highly specialized realm of their distinct engineering practice. While the modern engineering enterprise shifts into the focal engineering venture, context and lifeworld will play an increasingly important role.

The specialization of engineering practice from out of the more concrete and practical realm of everyday being has eventuated in modern times in the conceptual and often rather theoretical discipline of modern engineering. However, before the dawn of the modern engineer, the premodern engineer worked in the more concrete nonsystemic lifeworld. The notion of system is really a modern notion. The discussions in this chapter hold rather strictly for the modern engineer and rather loosely for the premodern engineer. Premodern engineers were more integrated into their lifeworld than are modern engineers. Precisely the quantification of what was more qualitatively engaged with, the abstraction out of the wholeness of concrete life, has brought us the fruits of modern technoscience but also has left us a plethora of issues with which to deal. For instance, the disburdenments modern engineering has wrought have left us disengaged from formerly more integrated lives. Of course, many disengagements are welcomed precisely because of the disburdenments. I appreciate the freedom I gain by being able to drive to the store rather than walk. With the time saved I am free to do other things. But, again, the point that Borgmann makes is that we tend to fill up our free time with other disengaging distractions, and we need some engagement in our lives.

Consider the example mentioned in the introduction involving the act of picking up a telephone phone and calling a friend. Easy and safe, no need to travel from the woods into town or even across town and run the risk of getting mugged. No need to endure bad weather. But what's lost in this disburdenment? A greeting from a neighbor, a flock of pelicans flying in formation, a brisk wind that clears the cobwebs from the brain, the face-to-face encounters that disclose worlds with just a smile.

SYSTEM

We occasionally choose to be but usually just find ourselves in our lifeworlds in a systemic or a nonsystemic manner. In the premodern era, being in the lifeworld nonsystemically was more typical, because the notion of system, as I indicated before, had yet to come into its own. But today the system idea is everywhere. Habermas, however, opposes system to the lifeworld, by conceptualizing society as being in a state of tension between system and lifeworld. System is comprised of those technical structures, like administrative and economic mechanisms, which serve our technical interests of social order and material production, and especially today, security.[21] In my view, as I have mentioned, the realm of systems is still part of the lifeworld. Systemic being is just a specialized, quantitative, theoretical, or abstract way of being in the lifeworld. We generally find ourselves being in the lifeworld in an everyday nonsystemic but nevertheless socially integrated manner. We are being-toward, being-with, or being-for the others we encounter in our midst. Then from that primal way of being we might branch out into various systemic ways of being. We can engage in strategic action employing varieties of instrumental rationality. We can participate in theoretical discourses, like mathematics or physics, involving present-at-hand entities, about which we strive to concoct truths of a conceptual nature. Or we might just meditate on a mountain range in the holistic interconnection of its profusion of peaks, an experience that keeps us anchored in the nonsystemic lifeworld.

A system is an assemblage or combination of entities forming a complex or unitary whole. A mountain range certainly qualifies as a system, though a meditation upon that range probably doesn't. The systems approach to reality was a major breakthrough in the evolution of human understanding. There were always heaps of things here and there, but to distinguish wholes, which are greater than the sum of their parts from heaps, which are the simple sums of their parts, represented a significant human advance. Wholes are foundational to systems development, and system characterization evolves by virtue of describing the interaction among parts that constitute the whole. Worlds can be systems and so can atoms.

Almost anything can be a system or part of a system. A lot depends on my orientation to the situation. If I take "x" artfully as something to be with or play around with, then I am normally being nonsystemic. If I take "x" practically or pragmatically in terms of its efficiency, productivity, or

21. Deborah Kilgore, "In Defense of the Lifeworld," *Adult Education Quarterly* 49, no. 2 (Winter 1999): 122.

profitability, then I see it as connected to a system of other entities, and while engaging with it I am normally being systemic. And there is a whole range of ways to be in-between. Instrumental rationality, regimentation, objectification, and capitalization lie more toward the realm of systems. Intuitive reasonableness, subjectivity, and caprice lie more toward the realm of the nonsystemic lifeworld.

Systems come in two basic types: they can be discovered or they can be created. Discovered systems are natural systems. They are just there, forming part of the given aspect of reality. Though they do not require the human hand for them to come into being, they do require humans to represent, articulate, or describe them, say, via a work of art or a set of differential equations. Created systems are artificial systems. They require the human hand for them to come into being, and they require the human mind to order their parts into a functional connection of entities designed for a given purpose. They are technological systems. The purpose or *telos* of a natural system is given, whereas the purpose of a technological system is a product of human ingenuity.

Habermas does not stress systems in general, but rather the particular system called society, the systemic nature of society. By system, Habermas understands society from the point of view of action consequences. From this standpoint, society is conceived as a system of actions. These are maintained and stabilized through functional interconnections, where each action has a functional significance in terms of its contribution to the maintenance of the system, thus allowing the system—as a collection of subsystems—to become independent of the lifeworld. The economy and public administration, for instance, form two central subsystems in late capitalist societies. Habermas, again, wants to sharply distinguish system and lifeworld, while I see the former as a subset of the latter. In any event, Habermas claims that the systems aspect of society comes into view necessarily as a consequence of adopting an objective observer's point of view, while the lifeworld—which I would call the nonsystemic lifeworld—can come into focus only by entertaining caring and participative "I-thou" perspectives.[22]

I can dwell in my lifeworld in a nonsystemic way, for example, by just puttering in my garden. Or I can behold the wonder of a singular

22. See Buber, *I and Thou*.

experience, say, of this apple tree, just today coming into bloom. To incorporate this tree into a system of scientific interpretation employing, for example, the Cartesian methodology of abstraction/decomposition/reconstitution/control, seems to miss something vital. The wonder of it, we might say, would be usurped by a kind of systemic amazement. Of course, the systemic and scientific perspective on the tree is valuable and contributes to general knowledge that advances civilization as we know it. Singularity is sacrificed, as it generally must be as the price of progress. Nonsystemic lifeworld dwelling is one thing, but if I look at my garden as the configuration of growing vegetables, "volunteers" perhaps, that I happened to discover growing over in a corner, I can take it as a natural system, which might have a function like providing lettuce for some summer salads. Or I can take that configuration of vegetables as a technological system, if I consciously focus my energies on weeding, fertilizing, watering, and cultivating my garden for maximum yield.

Within the realm of systems, then, ends and means, subjects and objects, are abstracted out of the flux of everyday life. But these elements do not just remain suspended in abstract system space. They are worked over, made into a variety of things and systems, connected to and combined with other systems, then they are reembedded within the lifeworld in order to serve as useful entities for the sake of nonsystemic "socially active humanity."[23] That reembedding might be called colonization as suggested in Chapter 3. But that is not the end of the story. In turn, the ensuing human activity in the nonsystemic lifeworld serves to contextualize the systems that eventuate from the abstractions in the first place. Various actions and decisions call for certain systemic responses and not others. Decisions made as a result of the communicative action of humans in the lifeworld can be said to contextualize the realm of systems. Engineers in their systemic endeavors experience the contextualization process as the call of the need to which they respond in the enactment of their projects. Many layers of context, however, might muffle the call, and some modern engineers might reject the notion that the engineering project is really

23. This phrase is one I discovered many years ago upon first looking into Heidegger. I was always intrigued by the tone of condescension and sarcasm Heidegger projected onto scientific and commonsense thinking, as opposed to his brand of deep and profound meditative thinking. It made me realize that Heidegger, in spite of all his deficiencies, probably had an interesting sense of humor. See "The End of Philosophy and the Task of Thinking," in *On Time and Being*, trans. Joan Stambaugh (New York: Harper & Row, 1972), 58.

responding to a lifeworld of human need. Modern engineers sometimes get so separated from the contexts of their endeavors that they think of themselves as disembodied machines carrying out preassigned functions. We may see ourselves as automatons of our own design, or else servants of the machinery of capitalist enterprise.

Since premodern engineers were more naturally attuned to the contextualization process, for them the natural and environmental contexts were close at hand. Abstraction out of the concrete lifeworld was not so sharp. Quantification out of the qualitative lifeworld was not so precise. The complexity and rigor of modern societies, however, call for a more pervasive process of systematization, the process of making connections and combinations among various systems within the realm of systems. Such systematization, stemming from lifeworld decisions, and the connections of systems to aspects of the lifeworld are essential to the contextualization process. As Andrew Feenberg puts it: "The exorbitant role of systematization in modern societies is rooted in the success of the co-ordination media, money, power, and . . . technology, and the large-scale organizations they make possible."[24] But without integrations and human decisions in the nonsystemic lifeworld, in other words, contextualization, no systems can exist.

Everywhere we turn we see elements of system. The voice of context, more audible in premodern societies, requires a finely tuned ear in contemporary times. In addition, once that voice is tuned in, Herculean effort is often needed to resist the temptation to integrate its message into the process of engineering. What should remain a social integration, at home in the nonsystemic lifeworld, is all too easily made into a system integration. This is an instance of what Habermas calls colonization, which was investigated in Chapter 3. But no colonization exists without contextualization.

TECHNOLOGICAL SYSTEMS

Technological systems are certainly systems but not all systems are technological systems. Organizations and institutions, as well as artifacts and actors, are elements of technological systems. Groups of actors coalesce into networks of people who activate, maintain, manage, and control

24. See Feenberg, *Questioning Technology*, 206.

institutions and organizations. These institutions and organizations consist not just of people but also of artifacts—systems, organisms, devices, structures, networks—of uncountable variety. Technological systems are generally built up over long periods of time. The railroad system, for instance, was begun with the running of the first steam locomotive in 1804 in South Wales. The nineteenth century was the century of the railroad. Track was laid and lines were developed. Safety features like signals were added. Terminals and train stations were built. Eventually, the railroad system became a global phenomenon. None of the elements involved in the technological system of the railroad was naturally occurring. Each required explicit planning and action on the part of concerned humans who had a stake in the expansion of the railroad.

As mentioned earlier, the realm of systems can be decomposed into natural systems and artificial systems. Natural systems are what I call *a priori* systems comprised of other humans, with whom I share a world, and also ready-to-hand things. Artificial systems are technological systems, which I call *a posteriori* systems, comprised of systemic things, which include objectified humans and present-at-hand entities. Present-at-hand entities are put together into systems after being designed, developed, and produced. The humans involved in *a posteriori* systems are engaged in technological networks. Human and nonhuman entities are integrated into or enrolled in networks or technological systems for the sake of some function or other.[25] Functionalism is an essential feature of technological systems.

Technological systems have both physical and social parts. The physical parts of these systems are connections of objects of all sorts: lawnmowers, tractors, freight cars, televisions, pipelines, automobiles, computers, and so on. The social parts are connections of people and organizations that make possible the connections between physical objects: farmers, lawyers, bakers, and truck drivers might be the people involved, while grain elevators, refineries, factories, and supermarkets might be the organizations involved.[26] It is difficult to conceive of any facet of contemporary life that is completely untouched by a technological system. As another

25. Bruno Latour, "Where Are the Missing Masses? The Sociology of a Few Mundane Artifacts," in *Shaping Technology/Building Society: Studies in Sociotechnical Change,* ed. W. Bijker and J. Law (Cambridge, Mass.: MIT Press, 1992).

26. Ruth Schwartz Cowan, *A Social History of American Technology* (New York: Oxford University Press, 1997), 151.

example, an aircraft is an artifact which is inherently systemic and integrated into technological systems. It is tied to myriad physical and social systems that include inspection schedules and inspectors, maintenance activities, pilots, flight scheduling, air traffic control, fuel distribution systems, safety systems, computers of all kinds, flight attendants, and flight attendant schools.

Of course, technological systems contextualize the engineering project. They form the backdrop against which the engineering project traffics. Without an already existing technological system of computer networks and satellites, for instance, the newly engineered communication products that continually pour forth from the engineering enterprise would find no home. Again, most of the technological systems that currently contextualize the modern engineering enterprise were only dimly perceived, or were really nonexistent in the premodern era. Lifeworld decisions and judgments contextualized the premodern engineering endeavor directly with little influence from the realm of systems or the realm of technological systems.

THE ENGINEERING PROJECT

In modern times, the engineering project is a particular kind of technological system. Engineers are actors. Tools of the trade as well as engineered products that stem from the engineering design process are artifacts. Engineering itself, the process proceeding in an orderly fashion, as well as the various professional societies that delimit and guide the engineering enterprise in performance of socially responsible action are the organizations. Though these elements of engineering as technological system are identifiable in the modern era, they were, again, less obvious in the premodern era. Actors, like the craftsperson or artisan—the engineers of the day—and artifacts, like the clay used by the potter, were surely recognizable, but the institutions in which engineers engineered were not. The distinction between lifeworld and technological system was not as sharply drawn then as it is today, since artifacts in the premodern era were more integrated into the lifeworld, and the lifeworld, as a result, had more of a contextualizing force. Ethical and aesthetic mediations, according to Feenberg, embellished the elementary technical object with additional dimensions of being, new secondary qualities, which helped to keep the artifact embedded in the lifeworld. Contextualization by the lifeworld was then more pronounced. "The ornamentation of artifacts and their investment with ethical meaning are integral to production in all traditional cultures. The choice of a type of stone or feather in the making of an arrow is motivated not only by sharpness and size, but also by various ritual considerations that yield an aesthetically and ethically expressive object."[27]

Within the premodern engineering endeavor, the artifacts that were incorporated into technological systems maintained their uniqueness and were not swallowed up by the system. Such artifacts were personalized more than they are today. My knife, for example, which I used for chopping, cutting, carving, throwing, and so on, was *my* knife. In modern times I can go into a lab and use any oscilloscope—owned by the company, not by me—to measure voltages in a circuit. Context, especially the particulars of a given time and place, the people and their character, for instance, had more of an impact on the technological system at issue then than is the case today. For the most part, invention in the premodern era produced only tools and not machines. Premodern engineers and their tools were more "one" with each other, whereas modern engineer and their machines are more separated. The traditional engineering endeavor was more tuned to and entangled within its lifeworld,[28] more contextualized by it. The modern engineering enterprise is more removed from its lifeworld, and thus more decontextualized.

CONCLUSIONS

We have looked at the increasingly abstract layers of lifeworld, system, and technological system as contexts of the project of engineering. Decisions made in the lifeworld contextualize the realm of systems, the realm of technological systems, and the engineering project itself. The immediate context of the engineering project, the realm of technological systems, includes, especially in modern times but less so in the traditional era, technicized social, economic, and political systems. These systems influence (contextualize) the engineering project in a variety of ways. Yet, the engineering

27. See Feenberg, *Questioning Technology,* 206.

28. Carl Mitcham, "Notes Toward a Philosophy of Meta-Technology," *Techne: Journal of the Society for Philosophy and Technology* 1, nos. 1–2 (Fall 1995): 3.

project is not only influenced by these systems, but also influences them. In fact, in general in modern times, technological systems tend to be increasingly engineered, and even if engineered artifacts therein are kept to a minimum, many such systems fall under the sway of the engineering metaphor. At stake here is the process of colonization, which was pursued in Chapter 3. The engineered aspects of these technological systems reveal that force of colonization.

PART III

THE FOCAL ENGINEERING VENTURE

SEVEN

PRODUCT

In this chapter, I arrive at the heart of my project. Having discussed the premodern and modern takes on the engineering project, it is now time to look at the focal engineering venture. In the focal engineering venture it is the *product* of the engineering project that stands out. Focal engineering asks whether these systems, services, structures, devices, organisms, and networks being let loose upon the planet are good products, and if so, in what sense? The hope is that some products will be not just functional but able to fulfill and engage our lives in a deep and meaningful manner.

Consider a product. For the moment it does not matter what product. Assume that the engineers who designed this product were fair, honest, and caring in all their deliberations and activities. Assume that the processes of engineering practice that resulted in this product were aimed at social justice, environmental sustainability, and the health and safety of end users. The product itself could still have a dissonant, disengaging, or deadening effect on its eventual end-user. Focally engineered products seek to counter those effects. Focal engineering is my proposal for the kind of engineering that concerns itself with bringing into the lifeworld products that disburden us without disengaging us.

Let us assume the latest model Gizmo Delux sedan has just rolled off the assembly line. It has great gas mileage. It is affordable. It is safe, comfortable, and efficient. Does it contribute to the good life? I buy one and on the way home almost immediately find myself in the middle of a traffic jam. I am depressed, anxious, out of balance, disengaged, lacking harmony. Of course, I cannot blame these feelings on my new Gizmo Delux. It is just one of the many products that has given us this world of disburdenment.

The point is that most engineered products are pretty neutral when it comes to the good, engaged, meaningful life. Do no harm: that's the aim of the modern engineering enterprise. Do not be unfair. Do not be unsafe. Do not pollute. But to actually seek to do good, to contribute to an enlivening and resonant life of engagement? That is what the focal engineering venture is after.

The contexts will play a greater role in the engineering of focal reality than they do for the modern engineering enterprise. The elements of those contexts, benefiting from the focal engineering effort, will resonate with each other. At least that's the aim of the focal engineering venture.

Focal engineering promotes products that are compliant with the precautionary principle. The precautionary principle says essentially that we should proceed with caution with regard to the implementation of any new or proposed technology, especially when the current scientific and technological knowledge about that technology is deemed to be incomplete. It is only common sense, argue its advocates, to avoid nasty surprises from new technologies.[1] As advocates of the precautionary principle, focal engineers would certainly avoid putting into the world products whose potential for harm is unknown, such as, for example, genetically engineering foods. They may really be acceptable in some circumstance and at some time, but a conversation of all concerned should be a prerequisite before any implementation takes place. Focal engineers, however, will take this principle one step further and seek to engineer products that are not just safe and healthy but are also harmonious with their worlds.

This chapter will explore the possible contributions of engineering to the expansion of contextually situated focal reality. If engineers do have a role to play in this venture, they cannot just ask *how* a product is to be made, like a premodern engineer, or *how* and *what,* like a modern engineer. They will have to also ask *why* a new or newly proposed product should be brought into being.

EXAMPLE

The Internet seems to be everywhere these days. In addition to the Internet, also on the rise are *Intranets.* They function much like the Internet

1. Kenneth R. Foster, "The Precautionary Principle—Common Sense or Environmental Extremism?" *IEEE Technology and Society Magazine* 21, no. 4 (Winter 2002/2003): 12

but are typically local in scope. They have access to the Internet but not vice versa. An Intranet is protected by a firewall that keeps out unauthorized visitors. Can an Intranet be a focal product? Maybe not, in and of itself, but maybe it can contribute to a focal practice.

A focal thing or practice "gathers the relations of its context and radiates into its surroundings and informs them."[2] It contributes to the harmony of the people involved with the product at issue and the context or world of this involvement. So, whether we are looking at a specific product or a specific practice employing that product, *product, person or persons,* and *world* should be brought into consonant focus. Such a product in light of that consonant focus, harmonized with end-user and lifeworld, would be considered a focally engineered product.

When we are talking about the focal engineering of products that can contribute to a focal reality, we must talk specific products in specific contexts. As Steven Goldman puts it, "Engineers [unlike scientists] have to wrestle with the refractoriness of the particular in its particularity."[3] In any particular focal product, context or world is always at stake, generating that product's particularity or at least the conditions that make that particularity possible. Computers in general, say, can never be focal or nonfocal in and of themselves. But certain types of computers, or the practices they serve, in certain situations, involving certain kinds of people, can be focal (or not). The specific case I want to look at is discussed by Michael Arnold and involves the use of an Intranet being made in a new housing development at Williams Bay in Melbourne, Australia.[4]

Arnold uses the idea of social capital in his explication, drawing on the work of Robert D. Putnam and others. "For Putnam, social capital is a compound composed of two major elements—networks of civic engagement and norms of generalized reciprocity, which together generate social trust."[5] In a well-networked community, people engage each other in relationships of general respect and reciprocity, people trust each other and openly share information, thoughts, and feelings. Social capital is high. The practice of communal engagement in a face-to-face sense is primal. The Intranet used by the people in the housing development at Williams

2. See Borgmann, *Technology,* 197.
3. Goldman, "Philosophy, Engineering, and Western Culture," 126.
4. Michael Arnold, "Intranets, Community, and Social Capital: The Case of Williams Bay," *Bulletin of Science, Technology & Society* 23, no. 2 (April 2003).
5. Ibid., 80.

Bay is intended to enhance their already established face-to-face engagement. The point seems to be that a focally integrated community should already be in place—or be on the way to being in place—before the Intranet is put into the mix. In other words, for the Intranet to be considered a focal product contributing to a focal practice, that practice should already be in place.

The Internet as a medium of mass communication seems less likely to be a focal product. People do meet on the Internet, but those meetings are seldom prefaced by a nonvirtual face-to-face engagement. Although surfing the Internet can be absorbing—wherein you lose yourself—it is not often engaging—wherein you find yourself. The Intranet as a communication device is the more likely candidate for a focal product. The Williams Bay residents are striving to develop social capital by virtue of their meetings and get-togethers in the real world. They hope to further expand that social capital focally through the use of a community Intranet that fosters civic engagement and community networking.

The focalness of a given product, such as the Intranet at Williams Bay, can be appreciated only through a narrative that lays out the context within which the focally engineered product comes into being. That story or narrative emerges out of a conversation to which all participants contribute. That means the end-users especially, as well as those who may not be users but who may be impacted by the products. That must include the people with the power as well as those without. The voice and the vision of the focal engineer should be prominent in that conversation.

THE CONVERSATION OF THE LIFEWORLD

One thing the Intranet example reveals is that people need to discuss the pros and cons of using a new technology. Could the residents of Williams Bay have gotten along without the Intranet? Sure, no doubt. But they and the developers of the community have been having ongoing conversations about the prospects of enhancing community engagement via the use of an Intranet. And the conclusion thus far—these conclusions are almost always contestable or at least tentative—is that, yes, the Intranet does expand social capital.

The Williams Bay conversations are generally about how to enhance community involvement and social capital, and not specifically about the employment of a specific technology. I am proposing that an ideal instance of focal engineering needs to engage all involved parties in constructing a narrative within the *conversation of the lifeworld* regarding any new or proposed technology. In order for a product or a design for a product to be considered a focally engineered product, it needs to be discussed by all interested parties, including the engineers involved in the creation of the technology at issue. Social justice requires the disenfranchised to also have a voice. How will this product enliven our lives? And so on. It is particularly the case that we engineers should look at the two sides of ourselves, "the side of ourselves we invest in the technology when we created it, and the side we now exercise in using it."[6] And, I would add, we should look at our role as citizens.

Landon Winner offers us three guiding maxims about how to focus lifeworld conversations:

1. No innovation without representation. All affected get a voice.
2. No engineering without political deliberation. And that would necessitate ethical and social deliberation as well. Who gets what and why are questions that need to be addressed.
3. No means without ends. Clearly established and meaningful goals need to be discussed and agreed upon before any design, development, production, or distribution gets under way.[7]

The conversation of the lifeworld that I envision could begin with Langdon Winner's maxims but it would have to incorporate more explicitly the focal dimension, which asks about what kind of world we want to contribute to in order that we might live more harmoniously. A major problem here is that realizations of the dreams of a better world through focal engineering tend to be thwarted by cost considerations. How to get down to the bottom line, to make these transformations to a new way of being an engineer, being a focal engineer, cost effective? Well, we might start by expanding the notion of "cost" to include not just dollars and cents, but also costs to the soul, to the spirit. Social capital must play a

6. Stephen L. Talbott, "Netfuture," February 25, 2003, 6. This is a newsletter dealing with technology and human responsibility published by the Nature Institute and is available at http://www.praxagora.com/stevet/netfuture/.

7. Langdon Winner, "Artifact/Ideas and Political Culture," *Whole Earth Review,* no. 73 (Winter 1991): 23–24.

role. Here is another spot where the conversation of the lifeworld needs to focus.

Conversation in the lifeworld regarding both the Intranet and the Internet has been rather minimal. The Williams Bay conversation has really been the exception to the rule. The Internet to most of us seemed to come out of nowhere and was generally perceived as a general good. We have viewed it as a wonderful tool to gather information and to reach out and touch someone halfway around the world. Why fret about something so positive? In fact there is some question about whether or not we actually could have been focal about the Internet. It did not seem to emerge explicitly as a network planned by specific human intentions and decisions arrived at within the conversation of the lifeworld. Was it, then, technically determined rather than socially constructed? The Internet has been found to be like many tools that just appear without a conversation of the lifeworld looking intently at what they might be for—in direct violation of Langdon Winner's maxim that we have "no means without ends." Inventions often come into being without unique purposes in mind, but once entrenched, users come up with appropriate uses. Edwin Schlossberg points out that the Internet itself was invented to allow scientists and scholars at universities to exchange information. "It was only when a software package was invented that made access easy and free that many people realized the opportunity that the Internet provided."[8]

Lifeworld conversation takes many shapes and involves many different characters and personalities, including nature. But as Steven Talbott maintains, we converse in order to become our true selves. Being-with and being-for the other, conversation is the explicit connection that discloses the already implicit connection inherent in my being my primal self, my true connected self. We cannot and do not want to predict or control the exact course of a conversation. Not, at least, if we are looking for a *good* conversation. As Talbott puts it, "Revelations and surprises lend our exchanges much of their savor. We don't want predictability; we want respect, meaning, and coherence. A satisfying conversation is neither rigidly programmed nor chaotic; somewhere between perfect order and total surprise we look for a creative tension, a progressive and mutual deepening of insight, a sense that we are getting somewhere worthwhile."[9]

8. Edwin Schlossberg, *Interactive Excellence* (New York: Ballantine, 1998), 14.

9. Stephen L. Talbott, "Ecological Conversation: Wildness, Anthropocentrism, and Deep Ecology," *Netfuture*, no. 127, January 10, 2002, 3.

Although the jinni is out of the bottle, perhaps a lifeworld conversation is still called for with respect to the Internet. All future Internet extensions, applications, refinements, and so on, can still be deliberated. And the focal engineer wants to be part of that deliberation. We recognize the Internet as a technology that is given now, for a growing number of people, as part of our everyday way of being. But alternatively we can take the situation in hand and begin to shape it, or continue to shape it, as we see fit.

WHEN IS ENOUGH ENOUGH?

One thing the prudent venture of focal engineering would initiate is a slowing down of the almost manic rush to push out "the next best thing." But immediately the idea of thoughtful considerations comes into conflict with the capitalist mandate to maximize profits in the minimum amount of time. Progress demands a relentless pushing forward into a utopian technological future. In this regard, we humans, with our full-steam-ahead attitudes, are driving and at the same time driven. But we are really nothing special, a lot like all successful species. Nothing special, except for one thing, just one small thing, as Bill McKibben says, "which the apostles of our technological future have overlooked. One small thing that actually does set us apart. What makes us unique is that we can restrain ourselves. We can decide not to do something that we are able to do. We can limit our desires. We can say, 'Enough.'"[10]

Just because we can do x, y, z does not mean we should. And here comes the ethics question again, What kind of world do we want to live in? A good world, to be sure. It seems to me that most thoughtful people would agree that that does not mean more and more stuff. Juliet Schor claims that we need a "politics of consumption" to lay out ideas and policies that can guide our consumption in a sagacious manner. Here again is where we must invoke the virtue of *phronesis,* the wisdom to see the big picture and do the right thing at the right time. Schor's proposal for a politics of consumption consists of seven basic elements:

1. A right to a decent standard of living
2. Quality of life rather than quantity of "stuff"
3. Ecologically sustainable consumption

10. Bill McKibben, *Enough* (New York: Henry Holt, 2003), 205.

4. Democratized consumption practices
5. A politics of retailing and the "cultural environment"
6. Exposure of commodity "fetishism"
7. A consumer movement and governmental policy[11]

The details of these elements need to be worked out in the conversation of the lifeworld and should involve the focal engineer's input. What does it mean, for instance, to expose commodity "fetishism"? A fetish is an object that is seen to embody magical powers. We often look at complex technologies this way. My computer is full of magical properties that I do not understand, and I hold it in awe because of that. But it needs to be demystified. A computer is really just a device, in the background of which are processes of design, prototyping, marketing, programming, production, testing, and so on. Everything we consume has to have been, in one way or another, produced. At the very least, it seems, a new politics of consumption must promote high standards by accounting for safety, labor, environment, social justice, and other conditions under which products are made. "This argument has been of great political importance in recent years, with public exposure of the so-called 'global sweatshop' in the apparel, footwear, and fashion industries. Companies fear their public images, and consumers appear willing to pay a little more for products when they know they have been produced responsibly. There are fruitful and essential linkages between production, consumption, and the environment that we should be making."[12]

Focal engineering aims to make those linkages explicit. And, along the lines of another of Schor's elements, we might ask, what is involved in expanding the quality of life? Focally engineered products aim at this expansion. If more and more "stuff" is not the answer, perhaps the answer is that less is more.

Affluence is always having more. Wealth is always having enough. Modern engineering contributes to affluence. Focal engineering will bring into the world products that will increase our wealth. Affluence is glamorous, Borgmann concedes, and "wealth in comparison is homely, homely in the sense of being plain and simple but homely also in allowing us to be at home in our world, intimate with its great things, and familiar with

11. Juliet Schor, "The New Politics of Consumption," *Boston Review*, 1999 (Summer): 10–12; available at http://bostonreview.mit.edu/BR24.3/schor.html/.

12. Ibid., 12.

our fellow human beings."[13] The wealth of having enough and knowing just when enough is enough can replace the affluence of having ever more disposable commodities.

STRUCTURES

Most structures are purely functional. We use them. The job gets done. We move on to the next thing. But some structures are not just functional but are also embellishing and inspirational to the human spirit. Focal structures, like bridges, like the Golden Gate, reflect their world as they radiate out into it. They gather world to them as world gathers them to it. What was heavy becomes light. What was dark becomes illuminated. A focal structure, lightened and lit up, opens up to its world. Focal engineering strives to bring structures into the world that will enliven and embellish the patterns that constitute places as lived worlds. According to Borgmann, "To open structures up is to give them an intelligible design, to adjust them to their location, and above all to invite humans to walk along, across, and through them, to rest and linger in them, and to allow people to comprehend the structures from various angles and at various levels."[14] Focal structures resonate with the worlds in which they are embedded. Humans who take up with these structures are engaged with them and enlivened in their worlds.

Architect Christopher Alexander discusses enlivening structures that need not be great but can actually be humble. Consider a simple courtyard. The structural engineers involved in the design and construction of a courtyard might be teamed with architects, homeowners, planners, and environmentalists. If this were a focally engineered project, focal engineers would promote on their team an enlivening place where people could sit under the stars, or sit in the shade and enjoy the sunny day. The structure and materials that comprise the courtyard must resonate with the patterns of human life-events that occur there, and the people who will use the product determine those patterns. Alexander suggests that people will walk in and out of a well-designed courtyard, passing through it every day, in the course of their normal lives. Eventually the courtyard should become a familiar and natural place to go and be a place that is

13. See Borgmann, *Technology*, 223–24.
14. See ibid., 243–44.

used. But, on the other hand, a badly designed courtyard—one with restricted access or intended as a destination in its own right—tends to stay unused.[15] Other features that make courtyards enlivening are an interesting view from inside the courtyard looking out, which makes it a comfortable place to be, smooth transition from inside to outside, via a porch or veranda, which makes it easier to go out more often, and a variety of paths in the courtyard. Gradually the courtyard becomes a harmonious and enlivening place to be.

But if walls enclose a courtyard leaving no opening to a larger world, the courtyard produces a claustrophobic sense and is a deadening place. If it has no porch or veranda as a space halfway between inside and outside, it does not invite people in. If the courtyard has no crossing paths and only one path leading out to it, people will not frequent it. "They hope to be there, but the lack of paths across the courtyard make it a dead and rarely visited place, which does not beckon them, and which instead tends to be filled with dead leaves, and forgotten plants."[16]

An enlivening product integrates smoothly with the patterns of human life-events that characterize any particular place, and such places are typically nodes in a web of other such places: products, people, patterns, and places in engaging accord. Focally engineered products, of course, may transform the patterns of human life-events, but they do so, or ought to do so, in such a way that the patterns are embellished. The enlivening patterns, enhanced by focal products, contribute a sense of stability and harmony to the lives of the human beings who use these products, as well as to the people who design and manufacture them.

DEVICES

Structures are devices but not all devices are structures. The essence of a device is its function. A device in Borgmann's sense of the *device paradigm* exists in such a way that its physical or mechanical presence tends to disappear or recede as its commodity aspect is made available.[17] Structures are devices, but they have a concrete physical presence that can be

15. Christopher Alexander, *The Timeless Way of Building* (New York: Oxford University Press, 1979), 109.

16. Ibid., 110.

17. See Borgmann, *Technology*, 40–48.

enlivening or deadening. The more deadening the structure, the less consequential its physical presence tends to be, and the closer it follows the device paradigm. When we confront an ugly bridge, all we want to do is get across it. Its functionality is all that matters. Its concrete physical presence becomes irrelevant.

As far as nonstructural devices are concerned, the focal engineer aims to engineer products that contribute to enlivening, engaging, and resonant life-events or focal practices. Electronic device engineering is a major concern of thousands of engineers today, most of whom would find it hard to relate to the notion of focal engineering as described here. They strive to get more and more computing power on smaller and smaller pieces of silicon real estate. But even though much of this work is done in "clean rooms" which are hermetically sealed from their surroundings, no engineer is an island. The chips they design are *for* something, something bigger. To shift into a focal engineering perspective, an engineer needs to inquire about the nature of this something. What is the device for, the device in which the chip will be embedded? How does it add beauty, peace, life, love, joy, to the world? In what way does it enhance human life-events or focal practices? Why should it be brought into the world? Why does it or why does it not support a sense of the good life in a convivial society? And replies to these questions will inevitably be debatable and open up other questions and other discourses.

PERSON/PRODUCT/WORLD ACCORD

The accord of person, product, and world refers to the harmonious manner in which the end user takes up with the product emerging from the focal engineering venture. At the same time, the end-user is enlivened in his or her lifeworld as a result of engaging with the product, and the product itself resonates with other things in its lifeworld. Among the reasons the modern engineering enterprise is incapable of attaining such accord are its inability to take the required extra time to accommodate this accord into product design because of economic constraints and its failure to exercise the virtue of *phroneses* because of its blindness to the relevance of context. The person/product accord, one of the aims of the modern engineering project in general and of the human-computer-interface (HCI) movement in particular, is necessary but not sufficient for focal engineering, which aims at a person/product/world accord. HCI seeks to have machines be more human, which is certainly a good idea. But humans, concomitantly, tend to become more like machines. Integrating world as context into the person/product accord helps to mitigate these tendencies. And that integration is precisely the aim of the practice of focal engineering. World *is* context and is intentionally brought to bear on the focally engineered product or person/product conjunct. The larger patterns must be allowed to speak. They are part and parcel of focal engineering. They add their share of dependable structure to the lifeworld.

While modern engineering may incorporate a larger pattern here and there, it does so only when social forces—acting from outside the modern engineering enterprise—demand attention, as is often the case with environmental constraints. With regard to modern engineering, because social and political interests have for the most part remained outside what engineering values, the profession lacks, as Stephen Johnston, Alison Lee, and Helen McGregor argue, "an adequate definition of human or community needs on which to base ethical judgments on its activity. The subordination by engineering science of all other discourses means that engineering has defined its practice as lying outside the context in which it occurs."[18]

Because it denies context, the modern engineering enterprise will never attain *phronesis*—practical wisdom. Recollection of context slows down focal engineering and makes it less progressive than modern engineering. But focal engineering is certainly more holistic than modern engineering. Admittedly, one of the constraints to a full enactment of focal engineering is the modern engineering quest to minimize "time to market." How does a company maintain its competitive edge? Must focal engineering look for ways to mitigate the determining power of market forces?

The focally engineered product, in invoking context, brings the ideas of the Good and the Beautiful back into the picture. As philosopher Ken Wilber has suggested, the Good and the Beautiful have been suppressed by the hegemony of the True in the modern era.[19] The modern Western world had divided the cultural spheres of morals, art, and science, which before the Renaissance had been intimately intertwined, but by the end of the eighteenth-century scientists, artists, and intellectuals were all specializing

18. Stephen Johnston, Alison Lee, and Helen McGregor, "Engineering as Captive Discourse," *SPT Journal* (http://scholar.lib.vt.edu/ejournals/STP/v1_n3n4/Johnston.html): 4.

19. Ken Wilber, interview by Mark Matousek, *Utne Reader,* July/August 1998, 50–56, 106–7.

and heading in different directions with little communication among themselves. This led to fragmentation, isolation, and alienation. As Wilber pointed out, "A very aggressive science, coupled with industrialization, was allowed to colonize and dominate the realms of morals and art. With the good and the beautiful removed from science, the only truth was materialism, which led us to our current disaster."[20]

Engineers engineer not just products but worlds as well. They create aspects of human lived worlds by the products they put into these worlds. And the focal engineer aims at a person/product/world accord. To thwart the growing dominance of the Bad, the Ugly, and the False, focal engineering by knowing how, what, and why—or at least by asking how, what, and why—tries to put in place products and practices that contribute to the Good, the Beautiful, and the True.

WHY QUESTIONS

While *how* and *what* questions remain important to focal engineering, *why* questions become pivotal. They ask for reasons, aims, intentions, goals, the *telos* of the focal engineering venture. Often they are assumed or vaguely expressed, like in the Strategic Defense Initiative (SDI) of the early 1980s. President Reagan insisted that SDI would render nuclear weapons impotent and obsolete. As Langdon Winner points out: "From that moment forward most of the debate, for and against, focused on the question: Will it work? Will the technologies function as planned? In the heated debates that followed, people seized upon the instrumental issues—issues about 'how'—as if they were the truly essential ones." Ends were taken for granted and not much discussed, while means were all that mattered. People began to talk about gadgetry and instrumental concerns, the wonders of technological possibility. They distanced themselves from asking *why* questions and from thinking about the goals of the project and the relationship between ends and technical means.[21]

Certainly, how and what questions are an integral part of the technoscientific project—how to solve these equations, what materials to use,

20. Ibid., 55.

21. Langdon Winner, "Engineering Ethics and Political Imagination," in *Broad and Narrow Interpretations of Philosophy of Technology*, ed. Paul T. Durbin (Dordrecht: Kluwer Academic Publishers, 1990), 61.

how to work within this or that constraint, what procedures to employ. We live in a world of rapidly advancing technology in which more and more sophisticated tools and techniques are being engineered every day. But we seem to have forgotten to ask *why* we do what we do. The U.S. military, for example, has increasingly powerful weaponry at its disposal and a great many situations across the planet seem to justify applying these weapons as rapidly as possible. Yet the debate over the notion of a just war never seems to get off the ground. Typically missing from our discussions about new technical means are deeply grounded reasons to guide our choices.

Consider the quest for the genetic engineering of the human species. Do we not all want optimal children? The means to achieve the kind of human we want are becoming more and more available. It is beginning to be possible to modify some human genetic structures. But why? Why is such a project needed? What and whose goals are to be served? Before continuing with such a project, one needs to explore the reasons.[22]

CONCLUSIONS

In a letter to the editor in the IEEE *Technology and Society Magazine,* Robert Brook takes Langdon Winner to task for insisting in an interview that engineers be contextually sensitive:

> Winner is unrealistic in his desire that engineers become politicians and policy mavens. It is not the purpose of the designer of a more efficient power supply or a class A amplifier to look for its ramifications in the area of social justice. The economic need is where engineering design starts. Engineers certainly cannot be held accountable for the unintended and unforeseen consequences of technological development—many of which turn out to benefit society.[23]

Brook is making a category error. Winner's interview is about focal engineering, not modern engineering, which does have the limitations and

22. Ibid., 60.
23. Robert Brook, letter to the editor, IEEE *Technology and Society Magazine,* Summer 1999, 4.

constraints that Brook ascribes to it. Indeed, Winner states his position quite clearly: "Technical change ought to be guided by principles of social justice, psychological harmony, and personal dignity, rather than the untrammeled pursuit of efficiency and profit."[24] The idea of harmony is at the heart of the focal engineering venture, while efficiency is intrinsic to the modern engineering enterprise. The "modern" and "focal" notions point toward different categories of engineering, the latter being more inclusive than the former. And though a given engineer may adopt different features of each category, to varying degrees at various times, the distinction should help to clarify issues, especially when focal engineering is advocated and is taken to be at odds with modern engineering.

I am not suggesting that modern engineering should be abandoned. It will persist, as premodern engineering has. Modern engineering will hopefully bring forth products that are user-friendly, that exemplify values of ideal modern engineering practice, including versatility, durability, simplicity, and stability. Nevertheless, the product that is well engineered and achieves these ideals, and that satisfies professional standards of safety, efficiency, and cost effectiveness may, in spite of all these wonderful aspects, still have a gravely deadening effect on the well-being of hapless users. The danger is not modern engineering per se, but rather the exclusiveness of such a practice. Without alternatives, the world will have to endure, as it in fact does today, many deadening or disengaging effects. To counter these effects, to promote things and products that have positive character, focal engineering is called for.

24. Langdon Winner, "Technology as 'Big Magic' and Other Myths," *IEEE Technology and Society Magazine,* Fall 1988, 4.

EIGHT

MATERIAL ETHICS

In this chapter I will look at an ethics of products brought forth into the world by the practicing engineer. A given product, as I indicated in the last chapter, may be equitably distributed, totally safe, and environmentally benign, yet it may still deaden or disengage us. Truly focal products contribute to the good and open up rather than close down the world. In this chapter I will make these ideas more explicit by employing a type of ethics called *material ethics,* which stems from the work of Albert Borgmann. Material ethics seeks to assess the focal-ness of products that aim to augment our shared lifeworld. The specific values I associate with the material ethics assessment are *engagement, enlivenment,* and *resonance.*

Focal products are not the same as products that have undergone human factors engineering. Kim Vincente defines human factors engineering as "the unique area of engineering that tailors the design of technology to people, rather than expecting people to adapt to technology."[1] There is a huge and important literature in the human factors area. Part of that area is taken up by the HCI (human-computer-interface) discipline, which is of growing concern in light of the proliferation of computers.

Generally speaking, the physical and psychological sides of the human person are major concerns of the human factors engineer. One of the things these engineers do is document the psychological aspects of people and the design techniques that can be used to create a fit with those aspects.[2] So, we expect that the human factors engineers will make sure the relationship of the end-user with the end-product is seamless to ensure that the user finds the product absorbing. But the focal product aims to

1. Kim Vincente, *The Human Factor* (New York: Routledge, 2004), 1.
2. Ibid., 90.

be *engaging* rather than *absorbing*. Absorption with my family sedan, for example, means I can drive the car without needing to explicitly think about it. But engagement with my sports car is a different story. To drive it properly, I need to be fully present with it. Absorption is a kind of disburdenment. I know that the aim of much of our engineered technology is to disburden us from onerous tasks, but Borgmann's point, again, is that a life of total disburdenment is a life of total disengagement. Focal engineering aims to engineer products that will be disburdening but at the same time engaging, enlivening, and resonant. Human factors engineering takes into account only part of the relationship between user and product. Focal engineering takes into account every aspect of the three-way relationship of user, product, and lifeworld.

How do we make lifeworld more explicitly a part of our deliberations regarding the ethical assessment of a product? World or lifeworld is context, and focal engineering looks at every particular product as being embedded in its contexts first by asking about its orgins—its formal, material, and efficient causes—and second by asking about its final cause, its goal, its *telos*. These questions about causes and purposes, if kept on the table, animate the lifeworld and keep it alive in our ethical assessments.

A clear, honest, noncoercive, public discourse—a conversation of the lifeworld—will be required to assess the focally engineered product. A major roadblock to that discourse is that public policy orientations these days tend to favor a rather limited cost, risk, benefit analysis that downplays the role of ethics. Engineers, to distinguish themselves as focal engineers acting in concert with other lifeworld deliberators, must extend their purview beyond the merely pragmatic and efficient. They must develop a better appreciation of their role as informed citizens who through their works of engineering can contribute to an enhanced lifeworld.

In this chapter, I will discuss a methodology for assessing the focality of an engineered product. Material ethics assumes that the virtue ethics assessments associated with the personal ethics of the engineer and the process ethics assessments associated with the professional ethics of the engineering process have been made with positive results. Yet, virtue ethics and process ethics assessments are necessary but not sufficient. Material ethics, which is a kind of public policy ethics, is a different kind of engineering ethics. Unlike virtue ethics of the individual engineer and conceptual ethics of the professional engineer, material ethics requires the contributions not only of engineers, focal engineers, but also of a variety of other citizens.

FOCAL ENGINEERING IS LOCAL ENGINEERING

Having become almost a cliché, the expression "think globally, act locally" is seldom given a second thought, yet it has definite relevance to the focal engineer in the conversation of the lifeworld. Global concerns, of course, abound. There remain the global problems of hunger, pollution, overpopulation, poverty, and energy depletion, among others. Most of us feel helpless in the face of these problems. But hope dies hard, and many people are proposing local steps to be taken which, if compounded, would have global ramifications. If more people grew their own vegetables, for example, world hunger would no doubt diminish. Local issues can be addressed by focal engineering. As these local incidences proliferate, the global village starts to thrive.

Though global concerns are certainly important, the initial encounter with context is a local experience. At the center of the material ethics assessment procedure of the focally engineered product is the requirement that the outcome must be good, do good, or contribute to the good, within the context of the end-user's local involvements. Being in a lifeworld means being bound up with social and political contingencies as well as a wide range of other patterns of human life-events. Such a lifeworld, in which I dwell and to which I am bound, can be thought of as a local habitation or engineering ecology, similar to what Bonnie Nardi and Vicki O'Day refer to as information ecologies.[3]

> By this we mean settings in which we as individuals have an active role, a unique and valuable local perspective, and a say in what happens. For most of us, it means our workplaces, schools, homes, libraries, hospitals, community centers, churches, clubs, and civic organizations. For some of us, it means a wider sphere of influence. All of us have local habitations in which we can reflect on appropriate uses of technology in light of our local practices, goals, and values.[4]

The project of focal engineering aims to make the engineered world, the engineered ecology, engaging, enlivening, and resonant as a result of incorporation of this or that system, device, organism, service, structure,

3. Bonnie A. Nardi and Vicki L. O'Day, *Information Ecologies: Using Technology with Heart* (Cambridge, Mass.: MIT Press, 1999).

4. Ibid., ix.

or network. Even if the engineered is seemingly immaterial, for example, a virtual reality, it still has material consequences. These are of concern to material ethics. What kind of prospects does a focally engineered system or device need if it is to be considered good or at least contributing to the good? It must be able to provide enrichment and fullness of contextualized being, conceptual continuities, and community attunements. Since these things mean many different things to many different people, the point of departure for focal engineering is opening the dialogue.

CONSEQUENTIALISM

Material ethics is a form of consequentialism, which claims that the moral rightness of an act depends on its consequences. The consequences of choosing the focal product over the nonfocal product are a more harmonious lifeworld and a harmony in the day-to-day tasks that make these tasks eventful. Living in harmony and peace with ourselves and one another within our engineered world may not be possible in any absolute sense, but it can be a vision toward which we strive.

Is material ethics then just *utilitarianism?* No. Consequentialism is the broad category. Utilitarianism is one form of consequentialism. Material ethics—as I envision it—is another. Utilitarianism, which seeks the greatest happiness for the greatest number, is the most common form of consequentialism and, like Kantian ethics, lies more toward the *Moralität* end of the *Moralität/Sittlichkeit* spectrum, the realm of practical ethical judgments, within which a general notion of duty is paramount. Within it there is an insistence that the duties which bind agents are based on pure reason. Materials ethics lies more toward the *Sittlichkeit* end of the spectrum, the practical realm of moral being and conduct, the world of our everyday involvements in which we encounter reality directly, wherein we intuitively sense how to be in accord with "the ought" or "the should." What matters to material ethics is the harmony that ought to result from an end-user taking up with a focal product with both user and product embedded in a contextualizing lifeworld. At the root of its concerns, material ethics includes, in Borgmann's words, "a moral assessment that takes into account the concrete dailiness that channels our endeavors and aspirations."[5]

5. Borgmann, "Moral Assessment of Technology," 211.

ASSESSING THE PRODUCT

To assess a product using materials ethics, we need to know how well the interaction of product, end-user, and lifeworld realizes the values of engagement, enlivenment, and resonance. Engagement arises from the interaction between the end user and the product; enlivenment, from the interaction between the end user and the lifeworld as a consequence of using the product; resonance, from the interaction of the product and the lifeworld (Figure 4).

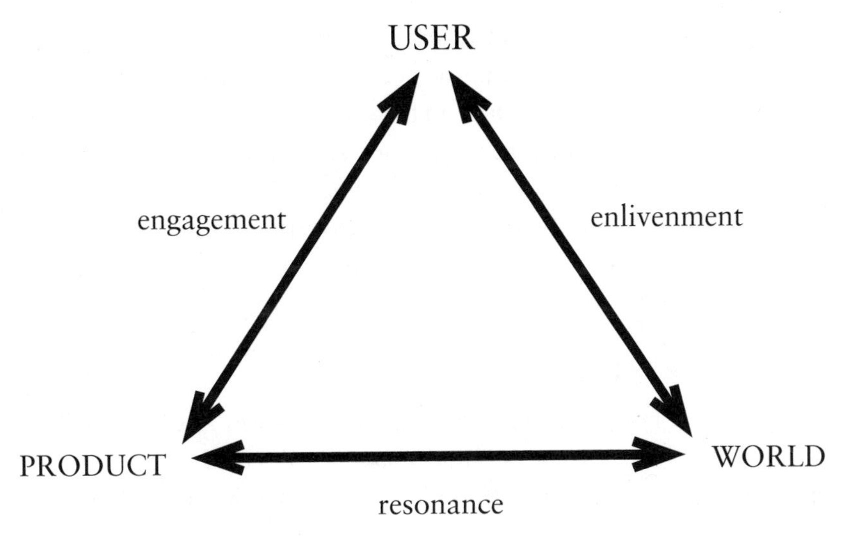

Figure 4. The Assessment Triangle

How might we go about making such an assessment? One way is the method I proposed earlier in the book for assessing the engineering process and the engineers involved in it. We can ask the participants in a discussion of the product in the lifeworld to assess the relationship between end-user and product (engagement), the relationship between end-user and lifeworld (enlivenment), and the relationship between product and lifeworld (resonance) on the now-familiar scale of −3 for very negative to +3 for very positive. As before, we will assign the average of their responses to the appropriate value function (J_{eg}, J_{el}, J_r) As before, the value functions are weighted (the γ_i terms). As before, assume initially that all three value functions are equally weighted ($\gamma_i = 1/3$). And as before, the numerical value of the weighted J function will indicate whether or not the materials ethics value associated with it has been achieved.

$$\text{By definition } J_m = \gamma_1 J_{eg} + \gamma_2 J_{el} + \gamma_3 J_r$$
$$\text{with } \gamma_1 + \gamma_2 + \gamma_3 = 1.0$$

Assume that as a result of conversations, we can gather a material ethics assessment of the RFID device, the example used in Chapters 2 and 5. The participants decided that the engagement of the end user with the RFID device would be rather minimal. The value of $J_{eg} = -2.0$ seemed appropriate. The enlivenment of the end user in his world also turned out to be minimal so $J_{el} = -2.0$ was deemed appropriate. The resonance of the RFID product and the human lifeworld was trickier to assess. This kind of resonance, impossible to measure, seemed to depend on the proponents of RFID making a case for how enlightened the world will be with a proliferation of RFIDs. Much discussion resulted in a value of $J_r = 1.0$. The total value function J_m is computed:

$$J_m = 1/3(-2) + 1/3(-2) + 1/3(+1) = -1.0$$

indicating a not very positive material ethics assessment. Again, this number is not the "answer" to the ethical question about whether or not the RFID device contributes to harmonious reality. But the number can provide a point of departure for further discussion.

CONSENSUS CONFERENCE MODEL

I assume that a conversation within an engineering group has done a virtue ethics assessment of the engineers involved in a given project, and I assume their assessment was positive, at least in terms of the personal virtues of fairness, honesty, and care. Furthermore, I assume that a conversation at the level of the profession has determined that the values of health and safety, environmental sustainability, and social justice have been honored by the engineering process. Nevertheless, if a company, for example, engineers a product about which there is some ambiguity regarding its environmental sustainability, then what? If the conversation at the level of the profession allows this product to be produced—perhaps giving it a green light because even though there was ambiguity regarding environmental sustainability, the other values of social justice and health and safety received high marks—then the ultimate burden of assessment falls upon the conversation of the lifeworld. Within that conversation the product is to be assessed primarily in terms of material ethics and the values of engagement, enlivenment, and resonance. However, within that conversation the virtue ethics and conceptual ethics assessments must also be reviewed. The conversation of the lifeworld is the Supreme Court of engineering ethics assessments. How might the conversation of the lifeworld unfold?

For example, people can just get together in a neighborhood or town meeting, talk things over, and decide on whether or not to install speed bumps. Take a vote and that's it. But what about more complex issues? Especially issues involving controversial engineered products? Genetically engineered or genetically modified (GM) foods are widely discussed today. Are the products or policies surrounding the products good or do they promote the good? Gathering, discussing, and voting seem to be necessary but not sufficient. Many layers of interpretation, for instance, support the notion of GM foods. The conversation could be more structured. Who, for instance, should be conversing? This is, I believe, the place where consensus conference models could come in handy. The consensus conference arrangement of public participation was first instituted by the Danish Board of Technology in 1987 and has been employed many times since, mostly in Europe.

A consensus conference can be defined as a method of technology assessment organized as a meeting between an expert panel and a lay panel consisting of concerned citizens.[6] The lay panel of citizens actually does the assessment after being informed by the experts. My suggestion is that both panels should include focal engineers. Focal engineers on the expert panel would, of course, be experts in the area of engineering from which the product springs. Focal engineers on the lay panel would be functioning in their role as citizens. A consensus conference is analogous to a jury

6. Johs Grundahl, "The Danish Consensus Conference Model," in *Public Participation in Science: The Role of Consesus Conferences in Europe*, ed. S. Joss and J. Durant (London: Science Museum, 1995); available at http://www.ucalgary.ca/~pubconf/Education/grundahl.htm.

process used in the courts. Sometimes they are called citizens' panels. Each panel consists of a representative cross-section of nine to fifteen citizens. These are people to be affected by the public policy under consideration. The citizens' panel is chosen by a steering committee whose membership includes only those who have no direct stake in the outcome of the policy recommendations.[7]

The expert panel is also chosen by the steering or planning committee, though the lay panel has some input as well. The steering/planning committee must be as impartial as possible. In a typical Danish consensus conference, the Danish Board of Technology, which is a quasi-independent state agency, sets up the steering committee.[8] Who controls the controllers is a question that sometimes comes up here. Striving for as much impartiality as we can on steering committees is the best we can do. Consensus conference recommendations are, as a matter of fact, seen as remarkably unbiased. Assuming a well-balanced steering committee is selected to oversee the consensus conference organization, what kind of constituency might that committee have? Typically it might include "an academic scientist, an industry researcher, a trade unionist, a representative of a public-interest group, and a project manager."[9] And I would, of course, strongly recommend the inclusion of focal engineers.

Focal engineers, then, should be on the steering committee as well as on the lay and expert panels. Focal engineers are, after all, citizens as well as engineers, and they have a special attunement to the good works that engineering can bring forth.

The theory behind Danish technology panels is that while experts can provide insight into the issues, mechanics, facts, possibilities, and problems associated with a particular technology, they do not have any better sense of what ought to be the case than the average citizen. In a democracy we the people should have the final say. Those of us whose lives are affected by an issue should have an effective voice in deciding how to

7. Bob Hudspith and Mike Kim, "Learning from a University-Cosponsored Regional Consensus Conference," *Bulletin of Science, Technology & Society* 22, no. 3 (June 2002): 232.

8. Phil Bereano, "Report on Danish 'Citizen Consensus Conference' on Genetically Engineered Foods, March 12–15, 1999," 2; available at http://www.loka.org/pages/DanishGeneFood.htm.

9. Richard E. Sclove, "Town Meetings on Technology," in *Science, Technology, and Democracy,* ed. Daniel Lee Kleinman (Albany: State University of New York Press, 2000), 35.

deal with it. The people who have to live with the results of the technology should be the judge about how to deal with the inevitable trade-offs. On these technology panels, people bring up for discussion their feelings, their values, their humanity, and the experiences of their everyday lives. This is precisely what is missing from most official dialogue about technical issues.[10]

The experts and the people should strive for consensus; however, the consensus sought here "is not the familiar Beltway political consensus, in which powerholder A trades favors with powerholder B, or where powerful interest groups forge lowest-common-denominator compromises at the expense of the rest of society." It is creative movement in the midst of inevitable differences and conflicts, seeking to achieve deeper and higher levels of common ground, often with unpredictable consequences.[11] Often in authentic dialogue there emerges an inspiration that cannot really be attributed to any particular speaker, but comes from the connections between participants that the dialogue promotes.

Now, the lay panel of the consensus conference, in my estimation, needs to expand its horizon. The ingredient I would like to insert into the mix of their deliberations is an assessment based on the ideas of material ethics. That would entail looking at any issue through the lens of the values of engagement, enlivenment, and resonance. I would think that a numerical value of 1.5 or above on the −3 to +3 scale mentioned above should be a requirement to give a project a green light. Of course, these and all such concerns should be open to discussion in the ongoing conversation of the lifeworld. In addition, if a consensus does not seem possible, I might want to include a constraint on variance. A lay panel with a high variance in their assessments could be assumed to have not yet finished their deliberations.

One more thing: the lay panel should also review the virtue ethics and process ethics assessments that are presumably done in-plant and at the professional level, respectively. Integrating the assessments of all three types of ethics—virtue ethics, process ethics, and material ethics—can result in an overall value function of the form

10. "Danish Citizen Technology Panels," 2, from a report of The Co-Intelligence Institute, available at http://www.co-intelligence.org/P-DanishTechPanels.html.

11. Ibid.

$$J = \alpha J_p + \beta J_v + \gamma J_m$$
$$\text{with } \alpha + \beta + \gamma = 1.0$$

and the Js are again restricted to a scale from −3 to +3. Since the notion of "at the professional level" is somewhat vague, we could have the process ethics assessment in terms of environmental sustainability, social justice, and health and safety done by professional engineers who are members of the expert and lay panels. These consensus conferences, in fact, might be optimal configurations for doing both process ethics and material ethics assessments. But that would require both the lay panel and expert panel to include professional engineers as well as focal engineers. They could, of course, be the same persons. By the expression *professional engineer* I mean a practicing engineer who is recognized as a professional by the society in which he or she functions.

CONCLUSIONS

Focal engineering and its assessment demand that before we decide on any suggested technological innovation we think seriously about it and talk about it. Reflection and discussion should precede letting loose upon the planet any new engineered product. This is not really anything new. Since our engineered world has transformed so rapidly, speed matters most, and many believe reflection is a luxury we can ill afford. But unless we make time for that reflection, the perception is that technological determinism prevails and the possibility of avoiding colonization of our lifeworlds appears slim. A slowing down and a step back are the minimal requirements.

There is a story of the Cheyenne Indian priest who consulted their most important god about whether the tribe should accept the new technology of horses when neighboring Comanches offered them to the tribe. He reported back to his people:

> If you have horses everything will be changed for you forever. You will have to move around a lot to find pasture for your horses. You will have to give up gardening and live by hunting and gathering, like the Comanches. And you will have to come out of your earth houses and live in tents. . . . You will have fights with

> other tribes, who will want your pasture lands or the places where you hunt. You will have to have real soldiers, who can protect the people. Think, before you decide.[12]

But, you may insist, the contemporary engineer is thinking all the time, seeking optimal solutions and "best practices." Contemporary engineers tend to be typically modern, which implies that they take on features of premodern engineering as well as a scientific perspective. Many projects, judgments, calculations, and decisions can be carried out in a premodern way. Design in the past was often intuitive. Design these days tends to be science based. Modern engineering explicitly employs science in the service of its methods and processes, particularly the design and manufacturing procedures, key ingredients in the production process. But the focus of the modern engineering enterprise is on means and methods. Ends and goals are usually considered to be outside the purview of the engineer. The discussion about ends and goals gets foreclosed because it is taken for granted that we all want to be disburdened and entertained.

A discussion of *aims* of the engineering project, as distinguished from the project itself and its methodology, brings modern engineering face to face with the possibilities of focal engineering. The venture can and should be directed toward the big problems of the day, like global warming, ozone depletion, declines in biodiversity, growing rates of resource depletion, and exponential population growth. Yet focal engineering seeks most earnestly to act locally, to embellish local ecologies with systems, services, devices, organisms, and structures whose prospects are good for advancing the engaged life. The best practice for a focal engineering enterprise might be not to bring forth such and such, but rather to decide against letting loose into the world another product that would lead to disengagement and dislocation.

To get to the point of seeing the crucial importance of focal engineering requires a fundamental reorientation of perspective, a shift in worldview, for both the engineering project and for our public discourse. In which public arenas might this best occur? Profit-driven corporations may be too entrenched in a bottom-line modern perspective, valuing only efficiency and productivity. The university, however, is touted as a place where

12. W. H. McNeill, "Goodbye to the Bison," *New York Review of Books,* April 27, 2000, 23.

students not only prepare for future careers, but also engage with contextualizing critiques of their lifeworld, putting it in perspective, and adopting a critical attitude toward it. However much it may have fallen from grace, the university is, I suggest, the most promising place to begin the conversation of focal engineering. For openers, without making drastic changes in curriculum, engineering instructors could initiate more *why* questions. Opening up the world in which a proposed device will function, *why* questions bring context to bear on the deliberations of professional engineering. How might the precise skills students learn in engineering fields be brought to bear upon the questions of connections between personal life and the social lifeworld? As Langdon Winner puts it: "Our moral obligations must now include a willingness to engage others in the difficult work of defining what the crucial choices are that confront technological society and how intelligently to confront them."[13] Clearly, the crucial choices for the ideal engineering project, the focal engineering venture, are choices about engineered products that are to be brought into the world.

The material ethics assessment is based on relating the values of engagement, enlivenment, and resonance to the constellation of product/user/world. The minimum requirement is the participation of the focal engineer in a conversation of the lifeworld about the product at issue. Ideally, such assessment should be done as soon as possible in the cycle of the product's design and development, at least before the product is brought forth into the world and made available for mass consumption.

I emphasized the consensus conference model of the conversation of the lifeworld because of its democratic, dialogical, and unbiased nature. The growing positive reputation of such conferences suggests that future technoscientific assessments can incorporate the voice of average citizens, including the disenfranchised, previously considered unable to understand the intricacies of the technoscientific phenomenon at issue. One doesn't really need to understand the details of, for instance, a radar system we may be assessing. But with expert input and intensive discussion, most average citizens who are serving on a lay panel can understand its uses and implications.

13. Langdon Winner, "Engineering Ethics and Political Imagination," in *Broad and Narrow Interpretations of Philosophy of Technology*, ed. Paul T. Durbin (Amsterdam: Kluwer Academic Publishers, 1990), 62.

NINE

BALANCE

The colonization and contextualization intrinsic to the engineering project will come into balance as the modern engineering enterprise shifts toward the focal engineering venture. The focal engineer contributes a new perspective from within contemporary technological culture, a perspective that will challenge our apparently accelerating plummet into a hypertechnological postmodern future. The hypermodern spirit seems to be on the verge of overwhelming us today, and our worldview seems to be uncritical of, indeed totally complicit in, that process. What we need is a balance of that spirit with a more critical worldview. What we witness, however, is a proliferation of all kinds of technologies, filling every corner of contemporary life. Some of it is of course useful, but it is hard to see how much of it contributes to advancing the quality of human life. The result, in Borgmann's palpably poetic expression, is "a suffocating overlay of disposable reality."[1]

To counter this descent, Borgmann recommends focal things and focal practices, and along Borgmannian lines, I am promoting a focally engineered reality. The focal engineering I am proposing is a specific kind of focal practice, the process of engineering itself as a focal practice, resulting in products that will either be candidates for focal things or will contribute to other focal practices. Such products can promote a commanding reality that balances the proliferation of disposable reality.

In this book I have been in pursuit of several states of balance that encompass the project of engineering. The balance of the old with the new, emotion with order, the general with the concrete, the one with the many, are all involved. But to be more specific, in this final chapter, I look at the

1. Borgmann, "Moral Significance of Material Culture," 299.

balanced *engineer* as the focal engineer, the balanced *engineering* project as the focal engineering venture, and the balanced *engineered* product as the focally engineered product.

Focal engineering will bring the colonizing forces of modern engineering and the contextualizing forces of premodern engineering into a state of equilibrium around the focally engineered *product.* These activities, within the consideration of the focally engineered product, become explicit concerns. Context is no longer side-stepped but becomes a crucial element in the focal engineering conversation of the lifeworld. Within that conversation, the material ethics assessment evolves, and harmony becomes the measure against which any proposed focal product stands. If the product does not yield a harmonious accord among end-user, product itself, and the lifeworld, then it cannot be called a focal product. If it does, then nothing stands in its way. If the assessment happened to reveal a harmonious accord after a product is designed and produced, then we are generally pleased. But the more thoroughly focal product is the one designed from the ground up to exhibit that harmonious accord.

Balance refers not only to product but also to the *process* of focal engineering, specifically the local/global balance that the focal engineering venture strives for. To be focal is primarily to be local. But we are deluged these days with claims of a globalized world and with positions favoring the essentiality of the globalization process. It connects us all, so the stories go, on a very high level, and it may be the only salvation for the future of the human species. Upon closer observation of those connections, however, they appear to be primarily economic in nature.

But, in point of fact, I fail to engage, in a full sense of the word and for the most part, with anyone I cannot meet face to face. There are exceptions. Feenberg mentions a medical support bulletin board devoted to people with Lou Gehrig's disease. The people on the board connect globally with others and share deep feelings about dependency, dying, and sexuality. These online networks certainly exhibit focality. Though the frankness in some of these discussions "may owe something to the anonymity of the online environment," generally speaking, the quality of these discussions and involvements are buttressed by face-to-face meetings at picnics and other gatherings.[2] And though the focal is generally focused in the local, the global nature of human connection will no doubt continue to

2. Feenberg, *Questioning Technology,* 192.

thrive and will probably remain essential to most future endeavors. What the focal engineering venture really seeks is not local over global involvements but rather a harmonious balance of local and global forces.

In addition, the idea of balance refers to the *person*, the focal engineer, who achieves a balance between the moods we find ourselves in and the understandings we develop: as *Befindlichkeit* and *Verstehen* called these *disposition* and *projection* in chapter four. Other kinds of balance may also be relevant. A plethora of key distinctions, then, emerges everywhere we turn, calling for balances not only in persons, but also in processes and products. If they are to be deemed focal, within the focal engineering venture, it will be necessary that the engineered, engineering, and the engineer achieve a state of balance. The first systematic concern of the chapter, as we move in the direction of a state of balance, will be with the person, the focal engineer.

THE ENGINEER

Like most people, focal engineers see the world in terms of both actuality and possibility. They are actual engineers, having graduated from this or that engineering school, working typically at some particular company, on some particular project. But their possibilities, both as engineers and as citizens, are more extensive than the particulars of their lives. Possibilities are what connect them to other people and things in their lifeworld. We might say that possibilities are com-possibilities. Modern engineers, as engineers, generally accept the actual givens in their lives as being the crucial aspect of their being engineers. But focal engineers seek to balance the facts of their being with the possibilities of who they can be as engineers. They tend to be reflective in their outlook. For example, a focal engineer might be concerned with questioning a procedure for its longer-term effects on health and safety, or trying to understand what it means to be an authentic engineer, or figuring out how to go about bringing into the world harmonious products. Many philosophers privilege the realm of the possible and maintain that it is prior, in the sense of fundamentality, to actual existing entities. Possibilities or com-possibilities are always already there and presupposed whenever any actual entity comes into view. Those possibilities can be viewed, in fact, as constitutive of that actuality. As Michael Gelvin puts it in his commentary on Heidegger, "There must be a mode beyond the limits of the actual, if there is to be an understanding and indeed a conceptual use of the actual existing entities. Possibility, whether articulated in terms of freedom (Kant), will (Schopenhauer), or the will to power (Nietzsche), alone can provide the perspective of the self necessary to explain the occurrence of the philosophical activity."[3] That perspective provides the framework for understanding and interpreting any given actual situation. Those givens may certainly be real but what do they mean? If we use the term "reality" to mean the sum total of all existing entities in the world, then actuality is "more real" than possibility. But possibility, argues Gelvin, is more *meaningful* than actuality.[4]

The meaning of their role as engineers is what focal engineers struggle with, even though that struggle may be backgrounded in their daily push to get the product out the door. There is always a question about whether that product is a good product, a focal product. Possibility, then, is primary in the involvements of the focal engineer, while actuality is more important in the involvements of the modern engineer. Focal engineers, of course, are not unaware of their actual situation but rather seek a balance between the actual and the possible. And this is no doubt also true for reflective modern engineers, although that reflection tends to be primal for focal engineers and secondary for modern engineers.

To enhance this analysis we can look at the Heideggerian ideas of *Befindlichkeit* and *Verstehen,* the state in which one finds oneself and the understanding associated with that state. *Befindlichkeit* and *Verstehen,* which Heidegger refers to as *existentiales* or fundamental features of our basic condition of being-in-the-world, indicate the bringing of awareness to the facts of our actuality and possibility. I find myself having certain dispositions, having been cast into a world I did not create. I keep shuffling through my possibilities and trying to make the best of an often ill-understood situation. The understandings I do arrive at are usually partial and difficult to maintain. As an engineer I may feel focal one day and completely unreflective the next. Still, focal engineers persist in alternating between experiencing actual reality and seeking understandings and possible interpretations of that reality.

What I have in mind here can be thought of as a *hermeneutic* circle. Hermeneutics is the practice of bringing to light interpretations and

3. Michael Gelven, *A Commentary on Heidegger's Being and Time,* revised ed. (DeKalb: Northern Illinois University Press, 1989), 79.

4. Ibid., 80.

meanings drawn from the world or context of our involvements. Peter-Paul Verbeek elaborates on this idea by employing Don Ihde's scheme of mediated perception, in which "technology" mediates between the "self" and the "world." Verbeek takes Ihde to task for suggesting that humans are already given as such and world is already given as such and in between we find artifacts. The more appropriate way to envision this, according to Verbeek's *postphenomenological* perspective, is that subject and object mutually constitute each other. "The relation between subject and object always already precedes the subject and the object themselves, which implies that the subject and object are mutually constituted in their interrelation."[5] For us, the engineer is always already implicated in the world. World conditions engineer, and engineer conditions world—to some degree at least. And how humans are and how world is are contingent upon the engineered products we let loose into the world. Focal engineers realize that they and their worlds are not just impacted by those products but that persons (both engineers and end-users), products, and world are, in Verbeek's terminology, mutually constituted in their interrelation. The hermeneutics involved in this conflux of interactions circle between person and world, between product and person, and between product and world. A balance in these hermeneutic circles might be characterized by the notion of harmonious accord.

How, then, should focal engineers be? Of course, I will assume focal engineers practice the virtues of care, honesty, and fairness in their engineering work. But also, in balancing out their actual and possible states of being, they balance out the factual way they find themselves immersed in this or that specific situation (*Befindlichkeit*) with understandings (*Verstehen*) and interpretations pertinent to that situation. Balanced and focused, they emanate a commanding presence in the face of the vicissitudes of everyday engineering life. They realize that the fruits of their labors will not just impact but will actually *constitute* ways of being. They stand their ground in making clear that their practice aims at health and safety, environmental sustainability, and social justice. And their focal engineering activities yield a product that will harmonize with end users and their world in an engaging, enlivening, and resonant manner.

5. Peter-Paul Verbeek, *What Things Do: Philosophical Reflections on Technology, Agency, and Design,* trans. Robert P. Crease (University Park: Pennsylvania State University Press, 2005), 129–30.

ENGINEERING

Focal engineering assumes that the modern engineering values have been pursued: health and safety, environmental sustainability, and social justice. These are essential but they are not sufficient. The harmony that the focal engineering venture aspires to takes the project of engineering to a new level. Not a higher level but, in a sense, a lower level. The focal engineering process focuses on the local domain, and it does so as a counter to globalization, to the global capitalist movement that has become hegemonic in contemporary society. *Balance* is probably a better word than counter, since the forces of globalization are ubiquitous and, so it appears, insurmountable.

Much, in fact, is being written these days about the wonders of globalization, the international bonding that shapes the political and economic relationships of almost every country on the planet. In his book *The Lexus and the Olive Tree,* Thomas Friedman says that globalization "can be incredibly empowering and incredibly coercive. It can democratize opportunity and democratize panic. It makes the whales bigger and the minnows stronger. It leaves you behind faster and faster, and it catches up to you faster and faster. While it is homogenizing cultures, it is also enabling people to share their unique individuality farther and wider."[6] So, globalization is ambiguous, that's for sure. And most of us are ambivalent about it. It invokes undecidability. Can we embrace the good stuff and shy away from the bad? Or do we have to either take it altogether or leave it? Or can we even leave it if we wanted to?

Whatever the case, the spread of the idea of the self-regulating market to every corner of the globe seems now to be a *fait accompli.*

Globalization is a decentralizing phenomenon. There is no center or structure of power and control. It erupts and has erupted everywhere at once. With no one in charge there is no visible hand, only the invisible hand of the market. Is globalization really just capitalism writ large? Does this mean that one cannot criticize globalization without also criticizing capitalism? Often, proponents as well as opponents of globalization choose to exaggerate the role of popular vehicles of globalization like NAFTA (North American Free Trade Agreement). They avoid any serious discussion about historical capitalism. As J. B. Foster maintains, "Radical

6. Thomas Friedman, *The Lexus and the Olive Tree* (New York: Anchor Books, 2000).

dissenters frequently single out the WTO, the IMF, the World Bank and multinational corporations—and even specific corporations like McDonalds—for criticism, while de-emphasizing the system, and its seemingly inexorable forces."[7]

One of the accepted if not acceptable consequences of globalization is the decline in the sovereignty of nation-states. The flow of capital eludes the constraints of local or national interests. Capital investors and multinational corporations are free to roam about the globe and are beholden to no nation-state or power structure. Their goal is endless and efficient production and accumulation of consumable commodities and a maximization of profits acquired through basic market exchanges. Labor, unlike capital, remains a local phenomenon, but it, like capital, serves the goals of productivity and efficiency. Also, labor is becoming increasingly devalued with the ascendancy of automation, leaving capital at the helm. Within this milieu of productivity and efficiency, capital colonizes the lifeworld. That is, if unfettered, globalized capital becomes the vehicle whereby, as Habermas claimed, the realm of system colonizes the lifeworld.

Nation-states, however, remain intact to varying degrees even though their sovereignty is diminished. They become rational players in the game of world culture. On the one hand, integration of nation-states can result in a global homogeneity because global concerns take precedent over local concerns and swamp out the uniqueness of any given particularity. On the other hand, that integration can bring forth a global heterogeneity with a mix of cultures yielding a global mosaic.

The mosaic versus melting pot idea used to be applied to the country but is now conceived globally. A key question is whether the global integration taking place today will turn out to be a *Gemeinschaft* (communal society) or a *Gesellschaft* (associational society). Because the former thrives on local engagements, the latter seems to be the case. In either case, however, there is no doubt that globalization has changed the essence of human interaction. As our culture pushes ever more strongly toward globalization, we sacrifice much that was near and dear. How can we recapture some of that local spirit without being romantic or nostalgic? For example, I find buying organic food locally at small natural food stores or at farmer's markets is more satisfying than shopping at the giant

7. J. B. Foster, "Monopoly Capital and the New Globalization," *Monthly Review*, January 2002, http://www.findarticles.com/p/articles/mi_m1132/is_8_53/ai8200 6/pg 2/.

supermarket chain. You get to know the customers and workers at small stores. It is more personal. Supporting local economy also feels more authentic. The global chains can provide me more instantaneous availability of foodstuffs than can the local providers. I cannot get organic and locally grown grapes at the farmer's market here in California in the winter. But just down the street at the supermarket I can buy grapes from Chile on the coldest and rainiest winter day. I prefer to just do without grapes in the winter. I feel more in tune with the rhythms of the local seasons.

As globalization pushes down across the boarders and on into South America and into the third world countries of Asia and Africa, we see the ill effects of organizations like NAFTA and the FTAA (Free Trade Area of the Americas). For one thing, the environment suffers due to market liberalization. Agricultural production suffers in the United States when we can buy food from the other side of the world at cheaper prices. Manufacturing, especially of the high-tech variety, endures a similar fate. These kinds of organizations are aiming to *liberalize* public services, which really means to *privatize* them. They seek, ultimately, to eliminate the commonweal. Public services are being viewed as inefficient, and by privatizing such services as education, health care, social security, water, postal delivery, and so on, we can save money and advance our quality of life. Of course, who makes out on these deals? Corporations to be sure. The poor people suffer as usual. And corporate privatization is on the rise in the United States and globally.

But focal engineering is local engineering. It aims to serve the public good, the commonweal. However, given that most engineering is a corporate concern, and most corporations are increasingly global, we must ask whether focal engineering is even possible. If it is not possible to keep context alive in the deliberations of the engineering project, then it might not be. But nowadays some concern for context is already mandated even by modern engineering ethics. The modern engineer is called upon to strive for environmental sustainability, health, safety, and social justice. The focal engineer is called upon to do the same, but other things as well. Contextual conversations need to be expanded, especially into areas of the possibility of a general harmony attributable to focal products. The smaller the corporation, the more likely it seems such conversations will be permitted and maybe even encouraged. Focal engineering, then, would seem to be at home in small start-ups and in companies that pride themselves on their human face. In any event, conversations are the key.

As an example, assume in my human-oriented company we are assigned the task of coming up with a proposal for engineering a movable walkway for a new airport in a developing third world country. Our nonfocal competitor submits a proposal that meets the specifications. Their design has minimal environmental impact, meeting all EPA standards. Its safety is guaranteed. As far as social justice is concerned, that's somewhat too vague and is left up to the politicians of the country. My focal engineering company has met all the technical and ethical standards that our competitor has. But we have an ongoing conversation with representatives of all concerned parties. Trips to the site were needed. There we discussed the walkway with airport users, owners, and workers. Our conversations were informal, but in general could be formalized along the lines of the Danish Consensus Conference model discussed in the previous chapter. How could our design harmonize with the hustle and bustle of the airport ambiance? We contemplated applying the principles of feng shui to make the movable walkway more resonant with the world of the airport. Will the world of the traveler be enlivened by using this product? Can they actually engage with the product, or will it just be something they use for its convenience and then forget about? Maybe that's all we want from the product. How might we make the experience memorable? Or is that even important or desirable? These are questions the focal engineer should address. We delved also into social justice issues that our competitor company glossed over. Will poor people be welcomed by this product? Of course they do not use airports much. But when they do, they should certainly feel welcomed. And what about the environment considered more deeply? Our competitors satisfied the minimal requirements, but we were interested in how the product/environment totality could actually contribute to social and environmental good, rather than having the product just do no harm. These kinds of concerns were more than just aesthetic overlays or optimal packaging of a given product. They were attempts to contextualize a product in the process of its realization. They were focalizations of our engineering practice.

However, will all this extra attention to the given engineering design context not be too costly? Will not the cheaper proposal that meets the specifications be the one selected? We think not. Here's why. Certainly the extra measures that focal engineering employs will cost more up front. But since we are a small company we have fewer CEOs, and the salary differential between labor and management is smaller for us. The April

14, 2003, issue of *Fortune* featured a story about the paychecks of corporate CEOs who pay themselves on average $7,452 per hour. Jim Hightower was incensed by this corporate greed. As he put it:

> These Thieves in Guccis are grabbing all they can for themselves at a time when their corporate performance stinks, shareholders are being stiffed, millions of workers are being dumped, pensioners are ripped off, unemployment is skyrocketing, college graduates are trading mortarboards for hairnets, and the general economy is rolling into a ditch. The stickiest-fingered Iraqi looter with a big cart and two mules has better ethics than our current corporate crowd.[8]

If we focal engineers have to compete with corporations with CEOs like this, we can afford to spend a bit more on the engineering process and still end up spending less overall than they do.

Another example concerns the possibility of focal agricultural engineering. Agricultural engineering uses scientific principles to design, test, and develop systems and equipment that deal with natural resources like water, soil, and energy for the production and processing of food, fiber, and feed. The efforts of agricultural engineering can be applied locally or globally. Most of this effort these days is at the global level. Most agricultural engineers seek employment with multinational corporations, which seek to feed the world by advancing efficiency and productivity of food, fiber, and feed products. Focal agricultural engineering tends to be local rather than global, but its overall aim is the same: to feed hungry people. Genetically modified (GM) food is becoming the silver bullet for many of the giant global corporations. The questioning of GM foods is expanding at the local level and goes hand-in-hand with promotion of local agriculture, which can benefit from the employment of focal agricultural engineering procedures. At the local level, farmers are going bankrupt—and this is true all around the world, not just in the United States—and rural life and local ecosystems are in decline, while the global food business is booming. The benefactors are a handful of giant agribusinesses, which focus mainly on bottom-line profits and ignore risks to human

8. Jim Hightower, *Hightower Lowdown,* June 2003, http://www.hightowerlowdown.org/.

health from chemical residues in global foods. The global food system consumes a major share of earth's resources and leaves in its wake a major amount of soil, air, and water pollution. And the expense of transport is an additional cost the consumer bears. The short-term gains that agribusiness accumulates are apparently grand enough to offset the long-term disasters waiting to happen, if indeed agribusiness even considers long-term consequences.

The turn from global to local agriculture is commendable. One obvious advantage is the lower transport costs. Shipping food a mile or two does not compare with shipping it a thousand. Local focal engineers seek to construct sustainable growing systems that leave the land in good shape for future generations. Organic farming is more labor intensive but it produces a product that resonates with the world of the consumer.

A prime example of local agriculture is a technique called *permaculture*. The focal engineering aspect of permaculture is located in the design and planning stages. As Bill Mollison tells us: "Permaculture is a design system for creating sustainable human environments."[9] It focuses on elements like animals, communications, plants, water, energy, and buildings. But permaculture is not about such elements per se, but rather about the relationships that are possible among them by virtue of how we place them in the landscape. A key idea here is to look at plants and animals in their total functionality. The inputs to one element come from the outputs of another. The needs of the system are met from within the system, unlike in the case of modern agriculture, which depends largely on external sources. Permaculture stresses diversity in crops, plants, trees, and animals. That diversity provides stability and helps us to prepare for environmental or social change. Permaculture also stresses small-scale, energy-efficient systems, which are intensive in that each element performs many functions and each important function is supported by many elements. Relative location is crucial. "To enable a design component (pond, woodlot, garden, windbreak, etc.) to function efficiently, *we must put it in the right place*."[10] Permaculture aims to rekindle the life of *place* in place of the soulless and abstract *space* modern engineering is typically thought in terms of.

9. Bill Mollison, with Reny Maislay, *Introduction to Permaculture* (Tyalgum, N.S.W., Australia: Tagari Publications, 1991), 1.

10. Ibid., 5.

What focal engineering seeks is an emphasis on local engagements, not as a replacement for global involvements, but rather as a complement to them. When engineering the movable walkway, for instance, we still use materials from our warehouses that are standards on the market and come from global sources. Some things are just not conveniently available locally, and we need to maintain that local/global balance. Within the traditional engineering endeavor, the local/global distinction was never a real issue. Within late modern engineering, the global aspect began to dominate. The focal engineering venture, stressing local involvements, in the context of the whole, seeks to provide answers to the question of balance.

THE ENGINEERED

I will assume that the focally engineered product has met all the requirements of the material ethics assessment and that the product contributes to harmonious life-events in a convivial society. What does the focal product denote or connote from a broader perspective? My contention is that the focal product is instrumental in balancing the forces of contextualization and colonization. Both forces have been extant to varying degrees throughout history. Contextualization is the effect that decisions made in the lifeworld have upon the engineering project. Colonization is the impact that the engineering project has on the human lifeworld.

I discussed triple colonization and colonization in general in Chapter 3. Technological systems are colonized by the modern engineering enterprise, systems by technological systems, and lifeworld by the realm of systems. Or, more directly, as is often the case, the engineering project colonizes the lifeworld directly.

Since colonization involves a process of decontextualization, if a balance is sought, then a contextualization is called for as a countermeasure. I should perhaps think of a reanimation of contextualization, rather than a re-creation of it, because contextualization has always been part and parcel of the engineering project, waxing in the premodern era, waning in the modern era. In contemporary times, if we seek to evolve a focal engineering practice, we will need to reestablish the connective linkages that constitute a healthy contextualization. Pure colonization would leave us suspended in an abstract, reductive, isolated—albeit efficient and productive—space. But pure colonization is never really possible, since some amount of contextualization is always already fleshing out the colonized and functionalized lifeworld.

Andrew Feenberg has proposed what he calls a Primary and a Secondary Instrumentalization.[11] The primary instrumentalization, associated with functionalization and colonization, involves the moments of *decontextualization* wherein the engineered product or the constitutive elements thereof are separated from their context; *reduction* wherein the primary qualities of the engineered product are separated from their secondary qualities; *autonomization* wherein the engineer or end user of the engineered product is separated from the product itself; and *positioning* wherein the subject situates itself strategically.[12] All these processes involve disconnection, separation, or taking apart. They are part and parcel of the general methodology of modern technoscientific activity, and they contribute to the colonization brought about by the impact of the modern engineering enterprise upon its lifeworld.

Still, acting in various ways as agents of their own interests, humans in the lifeworld are not generally just passive consumers. Only a philosophy of pure technological determinism would view the triple colonization process as a *fait accompli*. Social constructionists are champions of the secondary instrumentalization. They maintain that while the moments of the primary instrumentalization are enacted to varying degrees, there are also, often simultaneously, a series of contextualizing counter moments associated with the realization of any engineered product in its tangible human setting. Among the secondary instrumentalizations are the moments of *concretization* wherein synergisms are discovered between technologies and their various environments, *vocation* wherein humans bodily engage with their craft in a context of community, *mediation* wherein the ethical and aesthetic dimensions are brought more directly to bear on technical processes, and *initiative* or *collegiality* wherein the positioning process is collared to some degree as humans cooperatively appropriate available technologies.[13]

What might it mean that elements of the primary instrumentalization contribute to the colonization of the lifeworld by separating natural objects from their contexts, that is to say, by de-worlding them and reducing them to their primary qualities? Silicon, for example, is separated from the sand

11. See Feenberg, *Questioning Technology.*
12. Ibid., 203–5.
13. Ibid., 205–7.

in which it is found, then it is refined so that its conductivity is optimized. The silicon goes into the latest chip design. The chip becomes a microprocessor that serves as a controller for an assembly-line robot. Then the resulting electronic robot is programmed by an autonomous subject, eating donuts perhaps in a remotely situated control room. He controls the robot by optimally positioning himself with respect to the laws of mechanics that govern the robot's movements. The four moments of decontextualization, reductionism, autonomization, and positioning result in a robot adding value to an assembly line, making it more productive and efficient.

Robot, then, impacts the assembly line and transforms social relations. The wonders of our technological society must of course be celebrated. They lighten our loads in many ways. However, if only technological values dictate the shape of culture, then technological determinism can be said to prevail and the lifeworld will become narrowed and predictable, mechanized and monotonous. The effect of an engineered device impacting society can be compared to an open-loop control system. A setting, say, on an automatic clothes dryer is fixed at thirty minutes. After that time the machine automatically turns off. The clothes though may still be damp, or they may be scorched. Feedback would normally improve such a situation. In another feedback control system, for example, a thermostat-controlled central heating system, the actual temperature is sensed and turns the heater on or off in response to how close the actual temperature comes to the desired reference temperature. The secondary instrumentalization acts like a closing of the loop in a feedback control system, and social constructionism is its appropriate philosophy. The actual device of the robot must be combined with and connected to other elements of the overall system. The assembly line, the maintenance crew, the programmers, the products produced on the assembly line, these are all part of the system within which the robot is embedded. Ethical and aesthetic mediations provide our robot "with new secondary qualities that seamlessly embed it in its new social context."[14] Such mediations are contextualizing and act as counters to the reductionism inherent in the primary instrumentalization. The engineers who designed and produced the robot, as well as the end-users who take up with it, are not isolated subjects but counter the autonomization of the primary instrumentalization by working

14. Ibid., 206.

within the lifeworld traditions and practices that define their jobs, careers, or vocations. Employing various forms of initiative, end-users and engineers can invoke context that fleshes out their being-together and can lead to friendships and collegiality.

The four moments of initiative, vocation, mediation, and systematization integrate the robot into the lifeworld of involvements where its realization is contextualized. The engineered products that impact the lifeworld with their functionality are realized and these realizations are evaluated—this is where ethical assessments could occur within a conversation of the lifeworld—and feedback could be sent to the engineering project from the lifeworld, which may result in changes in the methods of design or modes of manufacturing. Ethical assessment could also take place in a more formal setting, for example, within a Danish Consensus Conference where the robot might be evaluated in terms of its service to the Good, how well it enhances quality of life and contributes to a more harmonious life of engagement, enlivenment, and resonance. In any event, lifeworld contextualizes the engineering project even as the engineering project impacts lifeworld.

Now, it is true that, via consumer attitude surveys and marketability studies, the modern engineering enterprise is influenced by the conversation of the lifeworld, the secondary instrumentalization is enacted, and the loop is closed to some degree. But within the modern engineering enterprise forces of contextualization are generally weak. To achieve a more authentic circle of mutual interpretation and dialogue between the engineering project and the lifeworld, to effect a balance, something like a shift from modern engineering to focal engineering is required.

Focal engineering incorporates the know-how of premodern engineering and the know-what stressed in modern engineering into an attitude that seeks to also know why. Focal engineering is inherently contextualizing because why questions point into context and bring context to bear on the multiple stages of the engineering enterprise. Focal engineering focuses on the public role of the practicing engineer. Public policy is made in the lifeworld, and the focal engineer plays an active role in the process of making policy about technological advances. Public policy, within the public arena, entails making decisions that encourage or request, and sometimes demand, that the engineering enterprise bring forth products that at least do no harm and ideally contribute to the common good of people in the lifeworld. Focal engineering pursues this ideal. It strives to bring into the world products that enliven and embellish the patterns of human life-events and eventful things populating an engaging lifeworld.

The product is the engineered. What about this focally engineered product? How does one gauge the good it is supposed to serve? The balance of the social construction *of* engineered products and the shaping of society *by* engineered products—contextualization and colonization—should constitute the backdrop against which the traffic of the focal engineering venture flows. A major issue here is that the ethics of the product, the engineered, is typically left to the end-user. But an authentic public policy would insist upon a focal engineering that aims to design, manufacture, and let loose upon the planet products that the end-user will find inherently tuned toward serving the common good. That is, enlivening and engaging products are human, sustainable, and of minimal risk. At the very least, a clear, honest, noncoercive, public discourse will be required for a genuine discussion about public policy and an enactment of an assessment process with the requirement that the outcome must be good, do good, or contribute to the good, within the context of the end user's involvements. Being in a lifeworld means being bound up with patterns of human life-events, with social and political contingencies.

THE FULCRUM

Focal engineering, then, effects a balance of the forces of colonization and contextualization. What might constitute the fulcrum for the balance that focal engineering promotes?

If we all shared a common religious belief or spiritual practice, that would do it. But we don't. We are a pluralistic society. Heterogeneity reigns. As a focal fulcrum, I propose the *Good.* Although ten people might have ten different notions of the nature of the Good, the conversation of the lifeworld could open up whatever common ground comes to light.

The Good is the ground, the foundation, upon which the edifice of the focal engineering venture is based. The Good is what the Greeks called the *hypokeimenon*—the foundation underlying all things. From it springs all that is worth having. The Good is the fulcrum, the point of balance, which is not just the origin but also the goal. It is for the sake of the Good that the focal engineer engineers the focally engineered product. For focal engineers to be good engineers, I suggested that they needed to be not only technically proficient but also practitioners of the virtues of fairness, honesty, and care. For the focal engineering process to be good, it needs to be aimed at the values of social justice, environmental sustainability, and health and safety. For the focally engineered product to be good, it needs to pay homage to values of engagement, enlivenment, and resonance. Again, the diversity of interpretations involved in these values calls for conversations, debates, negotiations. If all this transpires, and the conversations prove to be fruitful, then the Good is served. The result will be good products that contribute to the good life in a convivial society.

How do we get to be worthy people leading good lives who have the propensity to employ focal products in the service of a convivial society? About the good life, Aristotle said its highest goal is happiness. But the Greek word for what we translate as happiness, *eudaimonia,* meant something a little different to the Greeks. As Charles Guignon tells us:

> Where we usually think of happiness as a good feeling accompanying some activity or state, similar to pleasure, the Greeks regarded the feeling as only part of what constitutes happiness. For them, *eudaimonia* refers primarily to what Aristotle calls "living well and doing well," that is, living a life that is satisfying and worthwhile because it is full, abundant, and deserving of praise. This is why *eudaimonia* is often translated as "flourishing" or "thriving." What is at issue in this conception of happiness is not how one happens to feel at any moment, but the quality of one's life as a whole, with all its ties to the social world in which it unfolds.[15]

The good life is the life in pursuit of the good life. Ends and means are one, not two.

Charles Guignon believes that Aristotle's conception of the good life entailing character-building and acting in accord with virtues has had a deep and lasting impact on Western thought.[16] Though many believe, for example, that the practice of the virtue of love is necessary and sufficient for the good life, we cannot deny the technological nature of contemporary society and the fact that engineered products have saturated our

15. Charles Guignon, ed., *The Good Life* (Indianapolis, Ind.: Hackett, 1999), 22.
16. Ibid., 23.

world. In light of this and to augment the virtuous life, I contend that we also need the focal products that the focal engineering venture brings forth. They will augment and help, as Borgmann says, "to circumscribe a commonwealth of the good life."[17]

I have already indicated the kind of virtues that would be of benefit to the practicing focal engineer. This speaks of the *person as engineer*. What about the *engineer as person?* How to lead a good life is the question. What kind of character should citizens, including focal engineers, as end users of all kinds of engineered products, be in possession of in order to be worthy of taking up the task of determining the nature of the good life? That includes the determination of the focalness of engineered products. A minimal requirement would be a sound moral character. Of course, again, the heterogeneity of typical social life indicates a wealth of practices and positions that could be deemed appropriate and could be illustrated in narratives, stories, and conversations. The role of engineered products within these various practices and positions is also an important inquiry.

One interesting portrait of the good life comes from Richard Gula, a professor of moral theology. Because living the good life means living as a friend of what one considers a higher spiritual being or force, the moral and spiritual life must come together. The point of convergence is manifested in a person's character and virtue. Friendship is crucial. Even if God or gods are not part of one's spiritual path, we need to recognize a connection to something greater than ourselves, a higher power, or at least a whole of which we are a part, a One that contains the Many. Also, we need to connect to other people; and if we connect by establishing friendships, that, says Gula, is ideal. The good life is not about living in a world separate from ordinary life, but rather about living immersed in a life hopefully enriched by focally engineered products. The good life, as Gula maintains, is a life that we can live within the very activities and responsibilities that fill our days. The quest for the good life is about living virtuously in connection with others.[18]

In light of our connection to each other and to a higher source, we are both sacred and social beings. As sacred beings, we have dignity beyond our achievements. It is both a gift and a responsibility. As social beings, we live the good life by giving freely what we have received freely in

17. Borgmann, "Reply to My Critics," 366.
18. Richard M. Gula, *The Good Life* (New York: Paulist Press, 1999), 120–22.

order to create a community wherein everyone can flourish. Then Gula proceeds to recommend a set of virtues the practice of which will lead to the good life in a convivial society. Humility is one. It is a virtue that means we are realistic in trying to create a community where our sisters and brothers can gather. Such a community is characterized by equality, mutuality, and reciprocal giving and receiving for the sake of the well-being of all.[19] Gratitude is another virtue which entails giving what we can and only taking what is given. Self-esteem entails accepting and enjoying the blessings that befall us. Justice is a balancing kind of virtue in its own right which "reaches out to include those who can easily be forgotten—the lost, the least, the last." Other virtues Gula recommends include solidarity, fidelity, trust, hope and forgiveness. Forgiveness, for example, "is the virtue that makes it possible for us to live with one another without letting revenge and violence dominate our life together."[20]

Gula mentions one other virtue that ties all the others together. Hospitality. "It is the practice of paying attention to what is going on around us and then creating a welcoming space where we can experience new bonds of communion and live with all who are willing to put their gifts at the service of everyone else."[21] This may sound a bit too much like socialism or a Catholic religious community for the likes of some. But I think the idea here is that we must recognize and welcome our ties to the world and we must set up channels of communication that effectively allow us to interact with regard to issues of general concern, and that includes the discussion about the appropriateness or inappropriateness of particular engineered products. The conversation of the lifeworld needs to be made a part of our human involvement. As Vincent Miceli puts it in his book about the philosophy of Gabriel Marcel: "The thrust toward communication, communion and community represents a necessary facet of the interpersonal dialectic of the human situation."[22]

Jacques Derrida has a lot to say about hospitality. He speaks of *the* law of hospitality, which insists that the new arrival, the foreigner, the stranger, be offered an unconditional welcome. We ought to be yea-sayers and not yes-men, to welcome whomever or whatever might show up, as Derrida

19. Ibid., 125.
20. Ibid., 126.
21. Ibid., 126–27.
22. Vincent P. Miceli, *Ascent to Being: Gabriel Marcel's Philosophy of Communion* (New York: Desclee, 1965), 175.

insists "before any determination, before any anticipation, before any identification, whether or not it has to do with a foreigner, an immigrant, an invited guest, or an unexpected visitor, whether or not the new arrival is the citizen of another country, a human, an animal, or divine creature, a living or dead thing, male or female."[23]

Being-for the other is Levinas's expression for what it means to be convivial. *Love your neighbor as yourself* is what the New Testament says. *Love is all you need* said The Beatles. Iris Murdoch had a resonant take on the idea of the Good. She had discussed the indefinability of the idea of the Good, but thought there was something that could be said about it, something it had a relation to:

> Philosophers have often tried to discern such a relationship: Freedom, Reason, Happiness, Courage, History have recently been tried in the role. I do not find any of these candidates convincing. They seem to represent in each case the philosopher's admiration for some specialized aspect of human conduct which is much less than the whole of excellence and sometimes dubious in itself. I want to speak of what is perhaps the most obvious as well as the most ancient and traditional claimant, though one which is rarely mentioned by our contemporary philosophers, and that is Love.[24]

Murdoch did not want to conflate the Good and Love. The Good, she maintained, "is the magnetic centre toward which love naturally moves." Like Eros from Plato's *Symposium,* Love is capable of infinite degradation, "but when it is even partially refined it is the energy and passion of the soul in search for The Good, the force that joins us to Good and joins us to the world through The Good. Its existence is the unmistakable sign that we are spiritual creatures, attracted by excellence and made for The Good. It is a reflection of the warmth and light of the sun."[25]

The fulcrum, then, which I am naming *the Good* provides the foundation, the ground, for the balance of focal engineering. But the Good is also the aim, the goal, of the focal engineering venture. As ground we can

23. Jacques Derrida with Anne Dufourmantelle, *Of Hospitality,* trans. Rachel Bowlby (Stanford: Stanford University Press, 2000), 77.

24. Iris Murdoch, *The Sovereignty of Good* (London: Routledge & Kegan Paul, 1970), 99.

25. Ibid., 100.

perhaps view the Good as that which is attainable by the practice of the virtues, particularly, hospitality. As goal we can perhaps view the Good as inspiring the Love that aspires to the realm of the ideas, the home of the Good. Perhaps. But all these speculations might only raise more questions, issues for the conversation of the lifeworld to struggle with.

CONCLUSIONS

I have presented three kinds of ethics for three kinds of engineering. The premodern engineer practices the virtues suggested by virtue ethics. Modern engineering is guided by process ethics. The products of focal engineering are assessed by material ethics. Within the premodern engineering endeavor, engineer, engineering, engineered, and context were intimately bound up with each other. The project was a contextualized affair. Then as we shifted into the modern engineering enterprise, context became less crucial, and functionalization and colonization became the primary phenomena associated with the impact of engineering on the human lifeworld. In contemporary times the colonization process is continuing with a business as usual attitude as hypermodernism begins its reign, taking over from modernism but with refined and intensified procedures. Focal engineering is my suggestion for a way to implement what Borgmann calls postmodern realism, an alternative to hypermodern reality. Within the engineering of focal reality, the goal is to let loose upon the planet only those products that enliven, engage, and resonate. By recognizing and developing our common ground through conversations surrounding focal engineering, then seeing how we can affect the engineering project through our collective actions, we are able to amplify the forces of contextualization relevant to the engineering project. The result: a balance between the forces of colonization and contextualization.

It takes acting in the service of the Good to retain and maintain this balance. Hospitality in our everyday lives as well as being-for each other in the variety of ways it is done daily, solidify our connection to the Good and to each other as well. This web of association and relationship holds us up to feel the warmth and see the light of the sun.

INDEX